MATHEMATICAL
GAME
THEORY

MATHEMATICAL GAME THEORY

Ulrich Faigle

University of Cologne, Germany

World Scientific

NEW JERSEY · LONDON · SINGAPORE · BEIJING · SHANGHAI · HONG KONG · TAIPEI · CHENNAI · TOKYO

Published by

World Scientific Publishing Co. Pte. Ltd.
5 Toh Tuck Link, Singapore 596224
USA office: 27 Warren Street, Suite 401-402, Hackensack, NJ 07601
UK office: 57 Shelton Street, Covent Garden, London WC2H 9HE

Library of Congress Cataloging-in-Publication Data
Names: Faigle, Ulrich, author.
Title: Mathematical game theory / Ulrich Faigle, University of Cologne, Germany.
Description: New Jersey : World Scientific, [2022] | Includes bibliographical references and index.
Identifiers: LCCN 2021058963 | ISBN 9789811246692 (hardcover) |
    ISBN 9789811246708 (ebook) | ISBN 9789811246715 (ebook other)
Subjects: LCSH: Game theory.
Classification: LCC QA269 .F35 2022 | DDC 519.3--dc23/eng/20220125
LC record available at https://lccn.loc.gov/2021058963

British Library Cataloguing-in-Publication Data
A catalogue record for this book is available from the British Library.

For any available supplementary material, please visit
https://www.worldscientific.com/worldscibooks/10.1142/12540#t=suppl

Desk Editor: Liu Yumeng

Typeset by Stallion Press
Email: enquiries@stallionpress.com

Printed in Singapore

Contents

Preface

People have gambled and played games for thousands of years. Yet, only in the 17th century we see a serious attempt for a scientific approach to the subject. The combinatorial foundations of probability theory were developed by various mathematicians such as J. BERNOULLI[1] [4] as a means to understand games of chance (mostly involving rolls of dice) and to make conjectures according to mathematical principles.

Since then, game theory has grown into a wide field and appears at times quite removed from its combinatorial roots. The notion of a *game* has been broadened to encompass all kinds of human behavior and interactions of individuals or of groups and societies (see, *e.g.*, BERNE [3]). Much of current research studies humans in economic and social contexts and seeks to discover behavioral laws in analogy to physical laws.

The role of mathematics in this endeavor, however, has been quite limited so far. One major reason lies certainly in the fact that players in real life often behave differently than a simple mathematical model would predict. So seemingly paradoxical situations exist where people appear to contradict the straightforward analysis of the mathematical model builder. A famous such example is the *chain store paradox* of SELTEN[2] [41].

---

[1] J. BERNOULLI (1654–1705)
[2] R. SELTEN (1930–2016)

ix

This is not withstanding the ground breaking work of VON NEU-MANN and MORGENSTERN[3] [34], who have proposed an axiomatic approach to notions of utilities and rational behavior of the players of a game.

As interesting and worthwhile as research into laws that govern psychological, social or economic behavior of humans may be, the present *Mathematical Game Theory* is not about these aspects of game theory. In the center of our attention are mathematical models that may be useful for the analysis of game-theoretic situations. We are concerned with the mathematics of game-theoretic models but leave the question aside whether a particular model describes a particular situation in real life appropriately.

The mathematical analysis of a game-theoretic model treats objects neutrally. Elements and sets have no feelings *per se* and show no psychological behavior. They are neither generous nor cost conscious unless such features are built into the model as clearly formulated mathematical properties. The advantage of mathematical neutrality is substantial, however, because it allows us to embed the mathematical analysis into a much wider framework.

This introduction into mathematical game theory sees games being played on (possibly quite general) *systems*. A move of a game then correspond to a transition of a system from one state to another. Such an approach reveals a close connection with fundamental physical systems *via* the same underlying mathematics. Indeed, it is hoped that mathematical game theory may eventually play a role for real world games akin to the role of theoretical physics to real world physical systems.

The reader of this introductory text is expected to have knowledge in mathematics, perhaps at the level of a first course in linear algebra and real analysis. Nevertheless, the text will review relevant mathematical notions and properties and point to the literature for further details.

---

[3] J. VON NEUMANN (1903–1953), O. MORGENSTERN (1902–1977)

The reader is furthermore expected to read the text "actively". "Ex." marks not only an "example" but also an "exercise" that might deepen the understanding of the mathematical development.

The book is based on a one-term course on the subject the author has presented repeatedly at the University of Cologne to pre-master and master level students with an interest in applied mathematics, operations research and mathematical modelling.

It is dedicated to the memory of WALTER KERN.[4]

---

[4]W. KERN (1957–2021)

Part 1

Introduction

Chapter 1

Mathematical Models of the Real World

This introductory chapter discusses mathematical models, sketches the mathematical tools for their analysis, defines systems in general and systems of decisions in particular. Games are introduced from a general point of view and it is indicated how they may arise in combinatorial, economic, social, physical and other contexts.

1. Mathematical modelling

Mathematics is *the* powerful human instrument to analyze and to structure observations and to possibly discover natural "laws". These laws are logical principles that allow us not only to understand observed phenomena (*i.e.*, the so-called *real world*) but also to compute possible evolutions of current situations, and thus to attempt a "look into the future".

Why is that so? An answer to this question is difficult if not impossible. There is a wide-spread belief that *mathematics is the language of the universe*.[1] So everything can supposedly be captured by mathematics and all mathematical deductions reveal facts about the real world. I do not know whether this is true. Even if it were, one would have to be careful with real-world interpretations of mathematics, nonetheless. A simple example may illustrate the difficulty:

While apples on a tree are counted in terms of natural numbers, it would certainly be erroneous to conclude: *for every natural number*

---

[1]GALILEO GALILEI (1564–1642).

3

n, there exists a tree with n apples. In other words, when we use the set of nonnegative integers to describe the number of apples on a tree, our mathematical model will comprise mathematical objects that have no real counterparts.

Theoretically, one could try to get out of the apple dilemma by restricting the mathematical model to those numbers n that are realized by apple trees. But such a restricted model would be of no practical use as neither the set of such apple numbers n nor its specific algebraic structure is explicitly known. Indeed, while the sum $m + n$ of two natural numbers m and n is a natural number, it is not clear whether the existence of two apple trees with m resp. n apples guarantees the existence of an apple tree with $m + n$ apples.

In general, a mathematical model of a real-world situation is, alas, not necessarily guaranteed to be absolutely comprehensive. Mathematical conclusions are possibly only theoretical and may suggest objects and situations which do not exist in reality. One always has to double-check real-world interpretations of mathematical deductions and ask whether an interpretation is "reasonable" in the sense that it is commensurate with one's own personal experience.   ·

In the analysis of a game-theoretic situation, for example, one may want to take the psychology of individual players into account. A mathematical model of psychological behavior, however, is typically based on assumptions whose accuracy is unclear. Consequently, mathematically established results within such models must be interpreted with care, of course.

Moreover, similar to physical systems with a large number of particles (like molecules), game-theoretic systems with many agents (*e.g.*, traffic systems and economies) are too complex to analyze by following each of the many agents individually. Hence a practical approach will have to concentrate on "group behavior" and consider statistical parameters that average over individual numerical attributes.

Having cautioned the reader about the real-world interpretation of mathematical deductions, we will concentrate on mathematical models (and their mathematics) and leave the interpretation to the reader. Our emphasis is on *game-theoretic* models. So we should explain what we understand by this.

A *game* involves *players* that perform actions which make a given system go through a sequence of states. When the game ends, the system is in a state according to which the players receive *rewards* (or are charged with *costs* or whatever). Many game theorists think of a "player" as a humanoid, *i.e.*, a creature with human feelings, wishes and desires, and thus give it a human name.[2]

Elements of a *mathematical* model, however, do not have humanoid feelings *per se*. If they are to represent objects with wishes and desires, these wishes and desires must be explicitly formulated as mathematical optimization challenges with specified objective functions and restrictions. Therefore, we will try to be neutral and refer to "players" often just as *agents* with no specified sexual attributes. In particular, an agent will typically be an "it" rather than a "he" or a "she".

This terminological neutrality makes it clear that mathematical game theory comprises many more models than just those with human players. As we will see, many models of games, decisions, economics and social sciences have the same underlying mathematics as models of physics and informatics.

Note on continuous and differentiable functions. Real world phenomena are often modelled with continuous or even differentiable functions. However,

> • *There exists no practically feasible test for the continuity or the differentiability of a function!*

Continuity and differentiability, therefore, are *assumptions* of the model builder. These assumptions appear often reasonable and produce good results in applications. Moreover, they facilitate the mathematical analysis. Yet, their appropriateness cannot be proven

---

[2] *Alice* and *Bob* are quite popular choices.

by tests and experiments. The reader should be aware of the difference between a mathematical model and its physical origin.

Note on algorithms and computational complexity. Game-theoretic questions naturally call for mathematical computations within appropriate models. This will become clear in the present text, which also tries to exhibit important links to mathematical optimization theory. However, here is not the place to discuss specific mathematical optimization procedures *per se.* There is an abundance of classical mathematical literature on the latter, which can be consulted by the interested reader.

The question of the complexity of computations within particular game-theoretic models has attracted the interest of theoretical computer science and created the field of *algorithmic game theory,*[3] whose details would exceed the frame and aim of this text and are, therefore, not addressed either.

2. Mathematical preliminaries

The reader is assumed to have basic mathematical knowledge (at least at the level of an introductory course on linear algebra). Nevertheless, it is useful to review some of the mathematical terminology. Further basic facts are outlined in the Appendix.

2.1. *Functions and data representation*

A *function* $f : S \to W$ assigns elements $f(s)$ of a set W as values to the elements s of a set S. One way of looking at a function is to imagine a measuring device "f" which produces the result $f(s)$ upon the input s:

$$s \in S \; \longrightarrow \; \boxed{ f } \; \longrightarrow \; f(s) \in W.$$

We denote the collection of all W-valued functions with domain S as

$$W^S = \{f : S \to W\}$$

---

[3]See, *e.g.*, Nisan *et al.* [35].

and think of an element $f \in W^S$ also as a parameter vector whose coordinates f_s are indexed by the elements $s \in S$ and have values $f_s = f(s) \in W$.

There is a *dual* way of looking at this situation where the roles of the function f and the variable s are reversed. The dual viewpoint sees s as a *probe* which produces the value $f(s)$ when exposed to f:

$$f \; \longrightarrow \; \boxed{s} \; \longrightarrow \; f(s).$$

If S is small, the function f can be presented by a table which displays the total effect of f on S:

$$f \; \longleftrightarrow \;$$

s_1	s_2	s_3	$\ldots$	s_n
$f(s_1)$	$f(s_2)$	$f(s_3)$	$\ldots$	$f(s_n)$

The dual viewpoint would fix an element $s \in S$ and evaluate the effect of the measuring devices $f_1, \ldots, f_k$, for example, and thus represent an individual element $s \in S$ by a k-dimensional data table:

$$s \; \longleftrightarrow \;$$

f_1	f_2	f_3	$\ldots$	f_k
$f_1(s)$	$f_2(s)$	$f_3(s)$	$\ldots$	$f_k(s)$

The dual viewpoint is typically present when one tries to describe the state σ of an economic, social or physical system $\mathfrak{S}$ *via* the data values $f_1(\sigma)$, $f_2(\sigma)$, $\ldots$, $f_k(\sigma)$ of statistical measurements $f_1, \ldots, f_k$ with respect to k system characteristics:

$$\sigma \; \longleftrightarrow \; (f_1(\sigma), f_2(\sigma), \ldots, f_k(\sigma)).$$

The two viewpoints are logically equivalent. Indeed, the dual perspective sees the element $s \in S$ just like a function $\hat{s} : W^S \to W$ with values

$$\hat{s}(f) = f(s).$$

Also the first point of view is relevant for data representation. Consider, for example, a n-element set

$$N = \{i_1, \ldots, i_n\}.$$

In this context, a function $f : N \to \{0, 1\}$ may represent a subset S_f of N, *via* the identification

$$f \in \{0, 1\}^N \quad \longleftrightarrow \quad S_f = \{i \in N \mid f(i) = 1\} \subseteq N. \qquad (1)$$

REMARK 1.1. The vector f in (1) is the *incidence vector* of the subset S_f. Denoting by $\mathcal{N}$ the collection of all subsets of N and writing $\mathbf{2} = \{0, 1\}$, the context (1) establishes the correspondence

$$\mathbf{2}^N = \{0, 1\}^N \quad \longleftrightarrow \quad \mathcal{N} = \{S \subseteq N\}.$$

NOTA BENE. *$(0, 1)$-valued functions can also have other interpretations, of course. Information theory, for example, thinks of them as* bit vectors *and thus as carriers of information.*

An abstract function is a purely mathematical object with no physical meaning by itself. It obtains a concrete meaning only within the modelling context to which it refers.

Ex. 1.1. The so-called HEISENBERG and the SCHRÖDINGER pictures of quantum theory are dual to each other in the sense of the present section.[4]

Notation. When we think of a function $f : S \to W$ as a data representative, we think of f as a coordinate vector $f \in W^S$ with coordinate components $f_s = f(s)$ and also use the notation

$$f = (f_s \mid s \in S).$$

In the case $S = \{s_1, s_2, s_3 \ldots\}$, we may write

$$f = (f(s_1), f(s_2), \ldots) = (f_{s_1}, f_{s_2}, \ldots) = (f_s \mid s \in S)$$

Matrices. The *direct product* of two sets X and Y is the set of all ordered pairs (x, y) of elements $x \in X$ and $y \in Y$, *i.e.*,

$$X \times Y = \{(x, y) \mid x \in X, y \in Y\}.$$

---

[4] *cf.* Chapter 9.4.1.

A function $A : X \times Y \to W$ can be imagined as a matrix with rows labeled by the elements $x \in X$ and columns labeled by the elements $y \in Y$:

$$A = \begin{bmatrix} A_{x_1y_1} & A_{x_1y_2} & A_{x_1y_3} & \cdots \\ A_{x_2y_1} & A_{x_2y_2} & A_{x_3y_3} & \cdots \\ A_{x_3y_1} & A_{x_3y_2} & A_{x_3y_3} & \cdots \\ \vdots & \vdots & \vdots & \ddots \end{bmatrix}$$

The function values A_{xy} are the coefficients of A. We express this point of view in the shorthand notation

$$A = [A_{xy}] \in W^{X \times Y}.$$

The matrix form suggests to relate similar structures to A. The *transpose* of the matrix $A \in W^{X \times Y}$, for example, is the matrix

$$A^T = [A^T_{yx}] \in W^{Y \times X} \quad \text{with the coefficients} \quad A^T_{yx} = A_{xy}.$$

In the case $X = \{1, 2, \ldots, m\}$ and $Y = \{1, 2, \ldots, n\}$, one often simply writes

$$W^{m \times n} \cong W^{X \times Y}.$$

REMARK 1.2 (Row and column vectors). When one thinks of a coordinate vector $f \in W^X$ as a matrix having just f as the only column, one calls f a *column vector*. If f corresponds to a matrix with f as the only row, f is a *row vector*. So:

$$f^T \text{ row vector} \quad \Longleftrightarrow \quad f \text{ column vector}.$$

Graphs. A (*combinatorial*) *graph* $G = G(X)$ consists of a set X of *nodes* (or *vertices*) whose ordered pairs (x, y) of elements are viewed as arrows or (*directed*) *edges* between nodes:

$$\textcircled{x} \longrightarrow \textcircled{y} \quad (x, y \in X).$$

Denoting, as usual, the set of all real numbers by $\mathbb{R}$, an $\mathbb{R}$-*weighting* of G is an assignment of real number values a_{xx} to the nodes $x \in X$

and a_{xy} to the other edges (x, y) and hence corresponds to a matrix

$$A \in \mathbb{R}^{X \times X}$$

with X as its row and its column index set and coefficients $A_{xy} = a_{xy}$.

REMARK 1.3. Although logically equivalent to a matrix, a graph representation of data is often more intuitive in dynamic contexts.

A directed edge $e = (x, y)$ may, for example, represent a road along which one can travel from x to y in a traffic context. In another context, e could also indicate a possible transformation of x into y, *etc.*

The edge weight $a_e = a_{xy}$ could be the distance from x to y or the strength of an action exerted by x onto y, *etc.*

2.2. *Algebra of functions and matrices*

While the coefficients of data vectors or matrices can be quite varied (colors, sounds, configurations in games, *etc.*), we will typically deal with numerical data so that coordinate vectors have real numbers as their component values. Hence we deal with coordinate spaces of the type

$$\mathbb{R}^S = \{f : S \to \mathbb{R}\}.$$

Addition and scalar multiplication. The *sum* $f + g$ of two coordinate vectors $f, g \in \mathbb{R}^S$ is the vector of component sums $(f + g)_s = f_s + g_s$, *i.e.*,

$$f + g = (f_s + g_s | s \in S).$$

For any scalar $\lambda \in \mathbb{R}$, the *scalar product* λf multiplies each component of $f \in \mathbb{R}^S$ by λ:

$$\lambda f = (\lambda f_s | s \in S).$$

WARNING. *There are many — quite different — notions for "multiplication" operations with vectors.*

Products. The (*function*) *product* $f \bullet g$ of two vectors $f, g \in \mathbb{R}^S$ is the vector with the componentwise products, *i.e.*,

$$f \bullet g = (f_s g_s | s \in S).$$

In the special case of matrices $A, B \in \mathbb{R}^{X \times Y}$ the function product of A and B is called the HADAMARD[5] *product*

$$A \bullet B \in \mathbb{R}^{X \times Y} \quad (\text{with coefficients } (A \bullet B)_{xy} = A_{xy} B_{xy}).$$

WARNING. *The* HADAMARD *product is quite different than the standard matrix multiplication rule* (3) *below.*

Norm and inner products. Data vectors have typically only finitely many components. This means: One deals with parameter vectors $f \in \mathbb{R}^S$ where S is a finite set. Assuming S to be finite, the following notions are mathematically well-defined since they involve sums with only finitely many summands:

The (*euclidian*) *norm* of a vector $f \in \mathbb{R}^S$ is the parameter

$$\|f\| = \sqrt{\sum_{s \in S} |f_s|^2}.$$

The *inner product* of the vectors $f, g \in \mathbb{R}^S$ is the real number (!):

$$\langle f | g \rangle = \sum_{s \in S} f_s g_s = \sum_{s \in S} (f \bullet g)_s,$$

which allows us to express the norm as

$$\|f\| = \sqrt{\langle f | f \rangle}.$$

For finite matrices $A, B \in \mathbb{R}^{X \times Y}$, we have the completely analogous notions:

$$\langle A | B \rangle = \sum_{x \in X} \sum_{y \in Y} A_{xy} B_{xy} \quad \text{and} \quad \|A\|^2 = \langle A | A \rangle.$$

---

[5] J. HADAMARD (1865–1963).

Orthogonality. The idea of a norm is motivated by geometric considerations which suggest to interpret the norm $\|f\|$ as the *length* of the vector f. One says that $f, g \in \mathbb{R}^S$ are *orthogonal* if they satisfy the so-called *Theorem of* PYTHAGORAS, *i.e.*, the relation

$$\|f\|^2 + \|g\|^2 = \|f + g\|^2. \tag{2}$$

REMARK 1.4. Notice that the "Theorem of PYTHAGORAS" (2) is actually not a theorem but a definition (for the notion of orthogonality) in the context of the algebraic structure $\mathbb{R}^S$.

LEMMA 1.1 (Orthogonality). *Assuming S finite, one has for the coordinate vectors $f, g \in \mathbb{R}^S$:*

$$f \text{ and } g \text{ are orthogonal} \quad \Longleftrightarrow \quad \langle f|g \rangle = 0.$$

Proof. Straightforward exercise.

The standard matrix product. If we have matrices $A \in \mathbb{R}^{X \times Y}$ and $B \in \mathbb{R}^{Y \times Z}$, every row vector A_x of A is an element of $\mathbb{R}^Y$ and every column vector B^z of B is an element of $\mathbb{R}^Y$. So the corresponding inner products

$$\langle A_x | B^z \rangle = \sum_{y \in Y} A_{xy} B_{yz}$$

are well-defined. The matrix $AB \in \mathbb{R}^{X \times Z}$ of all those inner products is called the (*standard*) *product* of A and B.

The standard product of matrices $A \in \mathbb{R}^{X \times Y}$ and $B \in \mathbb{R}^{U \times Z}$ is ONLY declared in the case $U = Y$ and, then, defined as the matrix

$$C = AB \in \mathbb{R}^{X \times Z} \quad \text{with coefficients } C_{xz} = \sum_{y \in Y} A_{xy} B_{yz}. \tag{3}$$

If we think of $f, g \in \mathbb{R}^S$ as column vectors, then f^T defines a matrix with just one row f^T and g a matrix with just one column. With this understanding, we find

$$f^T g = \sum_{s \in S} f_s g_s = \langle f | g \rangle.$$

Ex. 1.2. If $f, g \in \mathbb{R}^n$ are column vectors, then $f^T g$ is formally a (1×1)-matrix with the single coefficient $\langle f | g \rangle$. This is to be clearly distinguished from the $(n \times n)$-matrix

$$fg^T = \begin{bmatrix} f_1 g_1 & f_1 g_2 & \cdots & f_1 g_n \\ f_2 g_1 & f_2 g_2 & \cdots & f_2 g_n \\ \vdots & \vdots & \ddots & \vdots \\ f_n g_1 & f_n g_2 & \cdots & f_n g_n \end{bmatrix} \in \mathbb{R}^{n \times n}.$$

Ex. 1.3 (Trace). The *trace* tr C of a matrix C is the sum of its diagonal elements. Show for the matrices $A, B \in \mathbb{R}^{X \times Y}$:

$$\langle A | B \rangle = \text{tr} \, (B^T A) = \sum_{y \in Y} (B^T A)_{yy}.$$

Linear maps. Recall that one may think of a matrix $A \in \mathbb{R}^{m \times n}$ as the parametrization of a linear map $f : \mathbb{R}^n \to \mathbb{R}^m$ that assigns to the n unit vectors $u_i \in \mathbb{R}^n$ the n column vectors of A:

$$f(u_i) = A u_i = \begin{bmatrix} A_{i1} \\ A_{i2} \\ \vdots \\ A_{im} \end{bmatrix} \in \mathbb{R}^m.$$

A general member of $\mathbb{R}^n$ is a linear combination of the n unit vectors u_i with scalar coefficients λ_i. *Linearity* of f means:

$$f(\lambda_i u_1 + \cdots + \lambda_m u_n) = \lambda_1 f(u_1) + \cdots + \lambda_n f(u_n) = \sum_{i=1}^{n} \lambda_i A u_i$$

and hence

$$f(x) = Ax \quad \text{for all (column vectors) } x \in \mathbb{R}^m.$$

REMARK 1.5. Where $I \in \mathbb{R}^{n \times n}$ is the $(n \times n)$ identity matrix (with column vectors u_i), one has

$$AI = A.$$

In contrast, assuming $A \in \mathbb{R}^{n \times n}$, the HADAMARD product yields a diagonal matrix:

$$A \bullet I = \begin{bmatrix} A_{11} & 0 & \cdots & 0 \\ 0 & A_{22} & \cdots & 0 \\ \vdots & \vdots & \ddots & \vdots \\ 0 & 0 & \cdots & A_{nn} \end{bmatrix}$$

Hilbert spaces. If S is an infinite set, both the sum $f + g$ and the HADAMARD product $f \bullet g$ are well-defined for any $f, g \in \mathbb{R}^S$. However, the notions of norms and inner products become vague because of the infinite sums in their definitions.

As a way around this difficulty, let us call a parameter vector $f \in \mathbb{R}^S$ a HILBERT[6] *vector* if

> (H$_1$) Only a countable number of coefficients f_s of f are non-zero.
>
> (H$_2$) $\|f\|^2 = \displaystyle\sum_{s \in S} |f_s|^2 < \infty.$

Let $\ell_2(S)$ denote the so-called HILBERT *space* of all HILBERT vectors in $\mathbb{R}^S$. $\ell_2(S)$ is a *vector space* over the scalar field $\mathbb{R}$ in the sense

> (V) $f, g \in \ell_2(S)$ and $\lambda \in \mathbb{R} \implies \lambda f + g \in \ell_2(S).$

As one knows form the algebra of series with at most countably many summands, norms and inner products are well-defined scalar numbers for all vectors in the HILBERT space $\ell_2(S)$.

---

[6]D. HILBERT (1862–1943).

While much of mathematical game theory can be developed in the more general setting of HILBERT spaces, we will be mostly concerned with finite-dimensional coordinate spaces in this introductory text.

Ex. 1.4. Show that all finite-dimensional real coordinate spaces are HILBERT spaces. In fact, one has:

$$\ell_s(S) = \mathbb{R}^S \quad \Longleftrightarrow \quad |S| < \infty.$$

2.3. *Numbers and algebra*

The set $\mathbb{R}$ of real numbers has an algebraic structure under the usual addition and multiplication rules for real numbers. $\mathbb{R}$ contains the set of natural numbers

$$\mathbb{N} = \{1, 2, \ldots, n, \ldots\}.$$

The computational rules of $\mathbb{R}$ may also be applied to $\mathbb{N}$ because sums and products of two natural numbers yield natural numbers.[7] Similar algebraic rules can be defined on other sets. We give two examples below.

REMARK 1.6. There is the philosophical issue whether "0" is a natural number, which corresponds to the question whether an entity can be a "set" when it is *empty*, *i.e.*, contains no element.[8] For clarification, we therefore employ the notation

$$\mathbb{N}_0 = \mathbb{N} \cup \{0\}$$

for the set of natural numbers *including* 0.

Complex numbers. There is no real number $r \in \mathbb{R}$ with the property $r^2 = -1$. To remedy this deficiency, one may introduce a new "number" i and do computations with it like it were a real number with the property

$$i^2 = -1.$$

---

[7]Though the same is not guaranteed for subtractions and divisions, of course.
[8]In Europe, for example, the idea and mathematical use of a number "zero" became customary not before the 13th century.

In doing so, one arrives at more general numbers of the form $z = a + ib$, with a and b being real numbers. The set

$$\mathbb{C} = \{a + ib \mid a, b \in \mathbb{R}\}$$

is the set of *complex numbers*. The special number

$$i = 0 + i \cdot 1$$

is the so-called *imaginary unit*. Using the algebraic rules of $\mathbb{R}$ and always keeping $i^2 = -1$ in mind, one can perform the usual computations with additions, subtractions, multiplications and divisions in $\mathbb{C}$.

Ex. 1.5. Assume that numbers $a, b, c, d \in \mathbb{R}$ are given such that

$$(a + ib)(c + id) = 1, \ i.e., \ c + id = (a + ib)^{-1} = \frac{1}{a + ib}.$$

Express the real numbers c and d in terms of the real numbers a and b.

A sober, neutral look at the matter reveals that we have imposed an algebraic structure on the set $\mathbb{R} \times \mathbb{R}$ of pairs (a, b) of real numbers with the computational rules

$$
\begin{aligned}
(a, b) + (c, d) &= (a + c, b + d) \\
(a, b) \cdot (c, d) &= (ac - bd, ad + bc).
\end{aligned}
\tag{4}
$$

If we identify the pair $(1, 0)$ with the real number 1 and the pair $(0, 1)$ with the imaginary unit i, we observe

$$(a, b) = a(1, 0) + b(0, 1) \in \mathbb{R} \times \mathbb{R} \quad \longleftrightarrow \quad a + ib \in \mathbb{C}.$$

It is straightforward to check that the rules (4) correspond precisely to the computational rules for complex numbers.

Another interesting view on complex numbers is offered by their representation as (2×2)-matrices:

$$a + \mathrm{i}b \in \mathbb{C} \longleftrightarrow a \begin{bmatrix} 1 & 0 \\ 0 & 1 \end{bmatrix} + b \begin{bmatrix} 0 & 1 \\ -1 & 0 \end{bmatrix} = \begin{bmatrix} a & b \\ -b & a \end{bmatrix} \in \mathbb{R}^{2 \times 2} \tag{5}$$

Ex. 1.6. Show: The sum and the product of two complex numbers is compatible with the matrix sum and (standard) matrix product of their representation of type (5).

We will see in Chapter 9 how the matrix representation (5) captures the basic idea of *interaction* among two agents and provides the fundamental link to the mathematical model underlying quantum theory.

For computational purposes, we retain:

Algebra in $\mathbb{C}$ follows the same rules as algebra in $\mathbb{R}$ with the additional rule

$$\mathrm{i}^2 = -1.$$

Binary algebra. Define an addition $\oplus$ and a multiplication $\otimes$ on the 2-element set $\mathcal{B} = \{0, 1\}$ according to the following tables:

$\oplus$	0	1
0	0	1
1	1	0

and

$\otimes$	0	1
0	0	0
1	0	1

Also in this *binary algebra*, division is possible in the sense that the equation

$$x \otimes y = 1$$

has a unique solution y "for every" $x \neq 0$.[9]

---

[9] There is actually only one such case: $y = x = 1$.

Vector algebra. Complex numbers allow us to define sums and products of vectors with complex coefficients in analogy with real sums and products.

The same is true for vectors with $(0,1)$-coefficients under the binary algebraic rules.[10]

REMARK. Are there clearly defined "correct" or "optimal" addition and multiplication rules on data structures that would reveal their real-world structure mathematically?

The answer is "no" in general. The imposed algebra is always a choice of the mathematical analyst — and not of "mother nature". It often requires care and ingenuity. Moreover, different algebraic setups may reveal different structural aspects and thus lead to additional insight.

2.4. *Probabilities, information and entropy*

Consider n mutually exclusive events $E_1, \ldots, E_n$, and expect that any one of these, say E_i, indeed occurs "with probability" $p_i = \Pr(E_i)$. Then the parameters p_i form a *probability distribution* $p \in \mathbb{R}^{\mathcal{E}}$ on the set $\mathcal{E} = \{E_1, \ldots, E_n\}$, *i.e.*, the p_i are nonnegative real numbers that sum up to 1:

$$p_1 + \cdots + p_n = 1 \quad \text{and} \quad p_1, \ldots, p_n \geq 0.$$

If we have furthermore a measuring or observation device f that produces the number f_i if E_i occurs, then these numbers have the *expected value*

$$\mu(f) = f_1 p_1 + \cdots + f_n p_n = \sum_{k=1}^{n} f_i p_i = \langle f | p \rangle. \tag{6}$$

In a game-theoretic context, a probability is often a *subjective* evaluation of the likelihood for an event to occur. The gambler,

---

[10]An application of binary algebra is the analysis of winning strategies for nim games in Section 2.6.

investor, or general player may not know in advance what the future will bring, but has more or less educated guesses on the likelihood of certain events. There is a close connection with the notion of *information*.

Intensity. We think of the *intensity* of an event E as a numerical parameter that is inversely proportional to its probability $p = \Pr(E)$ with which we expect its occurrence to be: the smaller p, the more intensely felt is an actual occurrence of E. For simplicity, let us take $1/p$ as our objective intensity measure.

REMARK 1.7 (FECHNER'S law). According to FECHNER,[11] the intensity of a physical stimulation is physiologically felt on a logarithmic scale. Well-known examples are the Richter scale for earthquakes or the decibel scale for the sound.

Following FECHNER, we *feel* the intensity of an event E that we expect with probability p on a logarithmic scale and hence according to a function of type

$$I_a(p) = \log_a(1/p) = -\log_a p, \qquad (7)$$

where $\log_a p$ is the logarithm of p relative to the basis $a > 0$ (see Ex. 1.7). In particular, the occurrence of an "impossible" event, which we expect with zero probability, has infinite intensity

$$I_a(0) = -\log_a 0 = +\infty.$$

NOTA BENE. *The mathematical intensity of an event depends only on the probability p with which it occurs — and not (!) on its interpretation within a modelling context or its "true nature" in a physical environment.*

---

[11]G.TH. FECHNER (1801–1887).

Ex. 1.7 (Logarithm). Recall from real analysis: For any given positive numbers $a, x > 0$, there is a unique number $y = \log_a x$ such that

$$x = a^y = a^{\log_a x},$$

where $e = 2.718281828...$ is EULER's[12] number, the notation $\ln x = \log_e x$ is commonly used. $\ln x$ is the so-called *natural logarithm* with the function derivative

$$(\ln x)' = 1/x \quad \text{for all } x > 0.$$

Two logarithm functions $\log_a x$ and $\log_b x$ differ just by a multiplicative constant. Indeed, one has

$$a^{\log_a x} = x = b^{\log_b x} = a^{(\log_b a)\log_b x}$$

and hence

$$\log_a x = (\log_b a) \cdot \log_b x \quad \text{for all } x > 0.$$

Information. In SHANNON's[13] [42] fundamental theory of information, the parameter

$$I_2(p) = -\log_2 p$$

is the quantity of information provided by an event E that occurs with probability p. Note that the probability value p can be regained from the information quantity $I_2(p)$:

$$p = 2^{\log_2 p} = 2^{-I_2(p)}.$$

This relationship shows that "probabilities" can be understood as numerical parameters that capture the amount of information (or lack of information) we have on the occurrence of events.

---

[12]L. EULER (1707–1783).
[13]C.E. SHANNON (1916–2001).

Entropy. The expected quantity of information provided by the family

$$\mathcal{E} = \{E_1, \dots, E_n\}$$

of events with the probability distribution $\pi = (p_1, \dots, p_n)$ is known as its *entropy*

$$H_2(\mathcal{E}) = H_2(\pi) = \sum_{k=1}^{n} p_k I_2(p_k) = -\sum_{k=1}^{n} p_k \log_2 p_k, \qquad (8)$$

where, by convention, one sets $0 \cdot \log_2 0 = 0$. Again, it should be noticed:

> $H_2(\mathcal{E})$ just depends on the parameter vector π — and *not* on a real-world interpretation of $\mathcal{E}$.

REMARK 1.8. Entropy is also a fundamental notion in thermodynamics, where it serves, for example, to define the temperature of a system.[14] Physicists prefer to work with base e rather than base 2 and thus with $\ln x$ instead of $\log_2 x$, *i.e.*, with the accordingly scaled entropy

$$H(\pi) = -\sum_{k=1}^{n} p_k \ln p_k = (\ln 2) \cdot H_2(\pi).$$

3. Systems

A *system* is a physical, economic, or other entity that is in a certain *state* at any given moment. Denoting by $\mathfrak{S}$ the collection of all possible states σ, we identify the system with $\mathfrak{S}$. This is, of course, a very abstract definition. In practice, one will have to describe the system states in a way that is suitable for a concrete mathematical analysis. To get a first idea of what is meant, let us look at some examples.

---

[14]Chapter 7 relates the notion of temperature also to game-theoretic activities.

Chess. A system arises from a game of chess as follows: A state of chess is a particular configuration C of the chess pieces on the chess board, together with the information which of the two players ("B" or "W") is to draw next. If $\mathfrak{C}$ is the collection of all possible chess configurations, a state could thus be described as a pair

$$\sigma = (C, p) \quad \text{with } C \in \mathfrak{C} \text{ and } p \in \{B, W\}.$$

In a similar way, a card game takes place in the context of a system whose states are the possible distributions of cards among the players together with the information which players are to move next.

Economies. The model of an *exchange economy* involves a set N of agents and a set $\mathcal{G}$ of certain specified goods. A *bundle* for agent $i \in N$ is a data vector

$$b = (b_G | G \in \mathcal{G}) \in \mathbb{R}^{\mathcal{G}},$$

where the component b_G indicates that the bundle b comprises b_G units of the good $G \in \mathcal{G}$. Denoting by $\mathcal{B}$ the set of all possible bundles, we can describe a state of the exchange economy by a data vector

$$\beta = (\beta_i \mid i \in N) \in \mathcal{B}^N$$

that specifies each agent i's particular bundle $\beta_i \in \mathcal{B}$.

Closely related is the description of the state of a general economy. One considers a set $\mathcal{E}$ of economic statistics E. Assuming that these statistics take numerical values ϵ_E at a given moment, the corresponding economic state is given by the data vector

$$\epsilon = (\epsilon_E \mid E \in \mathcal{E}) \in \mathbb{R}^{\mathcal{E}}$$

having the statistical values ϵ_E as its components.

Decisions. In a general *decision system* $\mathfrak{D}$, we are given a finite set

$$N = \{i_1, i_2, \ldots, i_n\}$$

of agents and assume that each agent $i \in N$ has to make a decision of a given type, *i.e.*, we assume that each i has to designate an element d_i in its individual "decision set" D_i. The joint decision of the members of N is then a n-dimensional data vector

$$d = (d_i | i \in N) \quad \text{with } n \text{ decision-valued components} \quad d_i \in D_i$$

and thus describes a *decision state* of the set N. In the context of game theory, decisions of agents often correspond to choices of strategies from certain strategy sets.

Decision systems are ubiquitous. In the context of a traffic situation, for example, N can be a set of persons who want to travel from individual starting points to individual destinations. Suppose that each person $i \in N$ selects a path P_i from a set $\mathcal{P}_i$ of possible paths in order to do so. Then a state of the associated traffic system is a definite selection π of paths by the members of the group N and thus a data vector with path-valued components:

$$\pi = (P_i | i \in N).$$

3.1. *Evolutions*

An *evolution* φ of a system $\mathfrak{S}$ over a time frame T is a function

$$\varphi : T \to \mathfrak{S}$$

with the interpretation: *The system $\mathfrak{S}$ is in the state $\varphi(t)$ at time t.* While the notion of "time" is a philosophically unclear issue, let us keep things simple and understand by a *time frame* just a set T of real numbers.

The time frames that are relevant in game-theoretic models are typically *discrete* in the sense that game-theoretic evolutions are observed at well-defined and well-separated time points t. So our time frames are of the type

$$T = \{t_0 < t_1 < t_2, \ldots < t_n < \cdots \}$$

Rather than speaking of the state $\sigma_n = \varphi(t_n)$ of a system at time $t = t_n$ under the evolution φ, it is often convenient to simply refer to the index n as the counter for the time elapsed. Hence an evolution φ corresponds to a sequence of states:

$$\varphi \quad \longleftrightarrow \quad \sigma_0 \sigma_1 \sigma_2 \ldots \sigma_n \ldots \quad (\sigma_n \in \mathfrak{S}).$$

4. Games

A *game* Γ involves a set N of *agents* (or *players*) and a system $\mathfrak{S}$ relative to which the game is played. A concrete game instance γ starts with some initial state $\sigma_0 \in \mathfrak{S}$ and consists of a sequence of *moves*, *i.e.*, of state transitions

$$\sigma_t \to \sigma_{t+1}$$

that are feasible according to the rules of Γ. After t steps, the system has evolved from state σ_0 into a state σ_t in a sequence of (feasible) moves

$$\sigma_0 \to \sigma_1 \to \cdots \to \sigma_{t-1} \to \sigma_t.$$

We refer to the associated sequence $\gamma_t = \sigma_0\sigma_1\cdots\sigma_{t-1}\sigma_t$ as the *stage* of Γ at time t and denote the set of all possible stages after t steps by

$$\Gamma_t = \{\gamma_t \mid \gamma_t \text{ is a possible stage of } \Gamma \text{ at time } t\}. \tag{9}$$

If the game instance γ ends in stage $\gamma_t = \sigma_0\sigma_1\cdots\sigma_t$, then σ_t is the *final* state of γ.

REMARK 1.9. It is important to note that there may be many finite state sequences γ in $\mathfrak{S}$ that are not feasible according to the rules of Γ and, therefore, are not stages of Γ.

Abstract games. The preceding informal discussion indicates how a game can be defined from an abstract point of view:

> - *A game Γ on a system $\mathfrak{S}$ is, by definition, a collection of finite state sequences $\gamma = \sigma_0\sigma_1\ldots\sigma_t$ with the property*
>
> $$\sigma_0\sigma_1\cdots\sigma_{t-1}\sigma_t \in \Gamma \quad \Longrightarrow \quad \sigma_0\sigma_1\cdots\sigma_{t-1} \in \Gamma.$$
>
> *The members $\gamma \in \Gamma$ are called the* stages *of Γ.*

Chess would thus be abstractly defined as the set of all possible finite sequences of legal chess moves. This set, however, is infinitely large and impossibly difficult to handle computationally.

In concrete practical situations, a game Γ is characterized by a set of *rules* that allow us to check whether a state sequence γ is feasible for Γ, *i.e.*, belongs to that potentially huge set Γ. The rules typically involve also a set N of *players* (or *agents*) that "somehow" influence the evolution of a game by exerting certain actions and making certain choices at subsequent points in time $t = 0, 1, 2, \ldots$.

Let us remain a bit vague on the precise mathematical meaning of "influence" at this point. It will become clear in special game contexts later.

In an instance of chess, for example, one knows which of the players is to move at a given time t. This player can then *move the system* deterministically from the current state σ_t into a next state σ_{t+1} according to the rules of chess. Many games, however, involve stochastic procedures (like rolling dice or shuffling cards) whose outcome is not known in advance and make it impossible for a player to select a desired subsequent state with certainty.

REMARK 1.10. When a game starts in a state σ_0 at time $t = 0$, it is often not clear in what stage γ it will end (or whether it ends at all).

Objectives and utilities. The players in a game may have certain *objectives* according to which they try to influence the evolution of a game. A rigorous mathematical model requires these objectives to be clearly formulated in mathematical terms, of course. A typical example of such objectives is a family U of *utility functions*

$$u_i : \Gamma \to \mathbb{R} \quad (i \in N)$$

which associate with each player $i \in N$ real numbers $u_i(\gamma) \in \mathbb{R}$ as its *utility value* once the stage $\gamma \in \Gamma$ is realized.

Its expected utility is, of course, of importance for the strategic decision of a player in a game. We illustrate this with an example in a betting context.

EX. 1.8. Consider a single player with a capital of 100 euros in a situation where a bet can be placed on the outcome of a $(0, 1)$-valued stochastic variable X with probabilities

$$p = \Pr\{X = 1\} \quad \text{and} \quad q = \Pr\{X = 0\} = 1 - p.$$

Assume:

- If the player invests f euros into the game and the event $\{X = 1\}$ occurs, the player will receive $2f$ euros. In the event $\{X = 0\}$ the investment f will be lost.

Question: What is the optimal investment amount f^* for the player?

To answer it, observe that the player's total portfolio after the bet is

$$x = x(f) = \begin{cases} 100 + f \text{ with probability } p \\ 100 - f \text{ with probability } q \end{cases}.$$

For the sake of the example, suppose that the player has a utility function $u(x)$ and wants to maximize the expected utility of x, that is the function

$$g(f) = p \cdot u(100 + f) + q \cdot u(100 - f).$$

Let us consider two scenarios:

(i) $u(x) = x$ and hence the expected utility

$$g(f) = p(100 + f) + q(100 - f)$$

with derivative

$$g'(f) = p - q = 2p - 1.$$

If $p < 1/2$, the derivative is negative and, therefore, $g(f)$ monotonically decreasing in f. Consequently $f^* = 0$ would be the best decision.

In the case $p > 1/2$, $g(f)$ is monotonically increasing and, therefore, the optimal utility is to be expected from the full investment $f^* = 100$.

(ii) $u(x) = \ln x$ and hence

$$g(f) = p \ln(100 + f) + q \ln(100 - f)$$

with the derivative

$$g'(f) = \frac{p}{100 + f} - \frac{q}{100 - f} = \frac{100(p - q) - f}{10000 + f^2} \quad (0 \le f \le 100).$$

In this case, $g(f)$ increases monotonically until $f = 100(p - q)$ and decreases monotonically afterwards. So the best investment choice is

$$f^* = 100(p - q) \quad \text{if } p \ge q.$$

If $p < q$, we have $100(p - q) < 0$. Hence $f^* = 0$ would be the best investment choice.

NOTA BENE. The player in Ex. 1.8 with utility function $u(x) = x$ risks a complete loss of the capital in the case $p > 1/2$ with probability q.

A player with utility function $u(x) = \ln x$ will never experience a complete loss of the capital.

Ex. 1.9. Analyze the betting problem in Ex. 1.8 for an investor with utility function $u(x) = x^2$.

REMARK 1.11 (Concavity). Utility functions which represent a *gain* are typically "concave", which intuitively means that the marginal utility gain is higher when the reference quantity is small than when it is big.

As an illustration, assume that $u : (0, \infty) \to \mathbb{R}$ is a differentiable utility function. Then the derivative $u'(x)$ represents the marginal utility value at x. u is concave if the derivative function

$$x \mapsto u'(x)$$

decreases monotonically with x.

The logarithm function $f(x) = \ln x$ has the strictly decreasing derivative $f'(x) = 1/x$ and is thus an (important) example of a concave utility.

Profit and cost. In a *profit game* the players i are assumed to aim at maximizing their utility u_i. In a *cost game* one tries to minimize one's utility to the best possible.

REMARK 1.12. The notions of profit and cost games are closely related: A profit game with the utilities u_i is formally equivalent to a cost game with utilities $c_i = -u_i$.

Terminology. A game with a set n of players is a so-called *n-person game*.[15] The particular case of 2-person games is fundamental, as will be seen later.

Decisions and strategies. In order to pursue its objective in an n-person game, an agent i may choose a *strategy* s_i from a set S_i of possible strategies. The joint strategic choice

$$s = (s_i | i \in N)$$

typically influences the evolution of the game. We illustrate the situation with a well-known game-theoretic puzzle:

Ex. 1.10 (Prisoner's dilemma). There are two agents A, B and the data matrix

$$U = \begin{bmatrix} (u_{11}^A, u_{11}^B) & (u_{12}^A, u_{12}^B) \\ (u_{21}^A, u_{21}^B) & (u_{22}^A, u_{22}^B) \end{bmatrix} = \begin{bmatrix} (7,7) & (1,9) \\ (9,1) & (3,3) \end{bmatrix}. \tag{10}$$

A and B play a game with these rules:

(1) A chooses a row i and B a column j of U.
(2) The choice (i, j) entails that A is "penalized" with the value u_{ij}^A and B with the value u_{ij}^B.

This 2-person game has an initial state σ_0 and four other possible states $(1, 1)$, $(1, 2)$, $(2, 1)$, $(2, 2)$, which correspond to the four coordinate positions of U.

The agents have to decide on strategies $i, j \in \{1, 2\}$. Their joint decision (i, j) will move the game from σ_0 into the final state $\sigma_1 = (i, j)$. So the game ends at time $t = 1$. The utility of player A is then the value u_{ij}^A. B has the utility value u_{ij}^B.

---

[15] Even if the players are not real "persons".

This game is usually understood as a cost game, *i.e.*, A and B aim at minimizing their utilities. What should A and B do optimally?

REMARK 1.13 (Prisoner's dilemma). The utility matrix U in (10) yields a version of the so-called *Prisoner's dilemma*, which is told as the story of two prisoners A and B who can either individually "confess" or "not confess" to the crime they are jointly accused of. Depending on their joint decision, they supposedly face prison terms as specified in U. Their "dilemma" is:

- No matter what decisions are taken, at least one of the prisoners will feel that he has taken the wrong decision in the end.

Part 2

2-Person Games

Chapter 2

Combinatorial Games

Games can always be understood as to involve two players that execute moves alternatingly. This aspect reveals a recursive character of games. The chapter takes a look at games that are guaranteed to end after a finite number of moves. Finite games are said to be combinatorial. Under the normal winning rule, combinatorial games have an algebraic structure and behave like generalized numbers. Game algebra allows one to explicitly compute winning strategies for nim games, for example.

1. Alternating players

Let Γ be a game that is played on a system $\mathfrak{S}$ and recall that Γ represents the collection of all possible stages in an abstract sense. Assume that a concrete instance of Γ starts with the initial state $\sigma_0 \in \mathfrak{S}$. Then we may imagine that the evolution of the game is caused by two "superplayers" that alternate with the following moves:

(1) The beginning player chooses a stage $\gamma_1 = \sigma_0\sigma_1 \in \Gamma$.

(2) The second player extends γ_1 to a stage $\gamma_2 = \sigma_0\sigma_1\sigma_2 \in \Gamma$.

(3) Now it is again the turn of the first player to realize the next feasible stage $\gamma_3 = \sigma_0\sigma_2\sigma_3 \in \Gamma$ and so on.

(4) The game stops if the player which would be next to move cannot find a feasible extension $\gamma_{t+1} \in \Gamma$ of the current stage γ_t.

This point of view allows us to interpret the evolution of a game as the evolution of a so-called *alternating* 2-person game. For such a game $\mathcal{A}$, we assume

---

(A_0) There is a set $\mathcal{G}$ and two players L and R and an initial element $G_0 \in \mathcal{G}$.

(A_1) For every $G \in \mathcal{G}$, there are subsets $G^L \subseteq \mathcal{G}$ and $G^R \subseteq \mathcal{G}$.

The two sets G^L and G^R in (A_1) are the sets of *options* of the respective players relative to G.

The rules of the alternating game $\mathcal{A}$ are:

(A_3) The beginning player chooses an option G_1 relative to G_0. Then the second player chooses an option G_2 relative to G_1. Now the first player may select an option G_3 relative to G_2 *and so on.*

(A_4) The game stops with G_t if the player whose turn it is has no option relative to G_t (*i.e.*, the corresponding option set is empty).

Ex. 2.1 (Chess). Chess is an obvious example of an alternating 2-person game. Its stopping rule (A_4) says that the game ends when a player's king has been taken (*"checkmate"*).

REMARK 2.1. While a chess game always starts with a move of the white player, notice that we have not specified whether L or R is the first player in the general definition of an altenating 2-person game. This lack of specification will offer the necessary flexibility in the recursive analysis of games below.

2. Recursiveness

An alternating 2-person game $\mathcal{A}$ as above has a *recursive* structure:

> (R) *A feasible move $G \to G'$ of a player reduces the current game to a new alternating 2-player game with initial element G'.*

To make this conceptually clear, we denote the options of the players L ("left") and R ("right") relative to G as

$$G = \{G_1^L, G_2^L, \ldots \mid G_1^R, G_2^R, \ldots\} \tag{11}$$

and think of G as the (recursive) description of a game that could possibly be reduced by L to a game G_i^L or by R to a game G_j^R, depending on whose turn it is to make a move.

3. Combinatorial games

Consider an alternating 2-person game in its recursive form (11):

$$G = \{G_1^L, G_2^L, \ldots \mid G_1^R, G_2^R, \ldots\}.$$

Denoting by $|G|$ the maximal number of subsequent moves that are possible in G, we say that G is a *combinatorial game* if

$$|G| < \infty,$$

i.e., if G is guaranteed to stop after a finite number of moves (no matter which player starts). Clearly, all the options G_i^L and G_j^R of G must then be combinatorial games as well:

$$|G| < \infty \quad \Longrightarrow \quad |G_i^L|, |G_j^R| \le |G| - 1 < \infty.$$

Ex. 2.2 (Chess). According to its standard rules, chess is *not* a combinatorial game because the players could move pieces back and forth and thus create a never ending sequence of moves. In practice, chess is played with an additional rule that ensures finiteness and thus makes it combinatorial (in the sense above). The use of a timing clock, for example, limits the the number of moves.

Ex. 2.3 (Nim). The *nim game* $G = G(N_1, \ldots, N_k)$ has two alternating players and starts with the initial configuration of a collection of k finite and pairwise disjoint sets $N_1, \ldots, N_k$. A move of a player is:

- Select one of these sets, say N_j, and remove one or more of the elements from N_j.

Clearly, one has

$$|G(N_1, \ldots, N_k)| \leq |N_1| + \ldots + |N_k| < \infty.$$

So nim is a combinatorial game.[a]

---
[a]A popular version of nim starts from four sets N_1, N_2, N_3, N_4 of pieces (pebbles or matches, *etc.*) with $|N_1| = 1, |N_2| = 3, |N_3| = 5$ and $|N_4| = 7$ elements.

Ex. 2.4 (Frogs). Having fixed numbers n and k, two frogs L and R sit n positions apart. A move of a frog consists in taking a leap of at least 1 but not more than k positions toward the other frog:

$$\text{\textcircled{L}} \rightarrow \bullet \; \bullet \; \bullet \cdots \bullet \; \bullet \bullet \leftarrow \text{\textcircled{R}}$$

The frogs are not allowed to jump over each other. Obviously, the game ends after at most n moves.

Remark 2.2. The game of frogs in Ex. 2.4 can be understood as a nim game with an additional move restriction. Initially, there is a set N with n elements (which correspond to the positions separating the frogs). A player must remove at least 1 but not more than k elements.

Creation of combinatorial games. The class $\mathfrak{R}$ of all combinatorial games can be created systematically. We first observe that there is exactly one combinatorial game G with $|G| = 0$, namely the game

$$O = \{\cdot \mid \cdot\}$$

in which no player has an option to move. Recall furthermore that all options G^L and G^R of a game G with $|G| = t$ must satisfy $|G^L| \leq t-1$ and $|G^R| \leq t - 1$. So we can imagine that $\mathfrak{R}$ is "created" in a never ending process from day to day:

DAY 0: The game $O = \{\cdot \mid \cdot\}$ is created and yields $\mathfrak{R}_0 = \{O\}$.

DAY 1: The games $\{O|\cdot\}, \{\cdot \mid O\}, \{O \mid O\}$ are created and one obtains the class

$$\mathfrak{R}_1 = \{O, \{O|\cdot\}, \{\cdot \mid O\}, \{O \mid O\}\}$$

of all combinatorial games G with $|G| \leq 1$.

DAY 2: The creation of the class $\mathfrak{R}_2$ of those combinatorial games with options in $\mathfrak{R}_1$ is completed. These include the games already in $\mathfrak{R}_1$ and the new games

$$\{\cdot|\{O|\cdot\}\}, \{\cdot \mid \{O|\cdot\}\}, \{\cdot \mid \{\cdot \mid O\}\} \ldots$$
$$\{O|\{O|\cdot\}\}, \{O \mid \{O|\cdot\}\}, \{O \mid \{\cdot \mid O\}\} \ldots$$
$$\{O, \{\cdot \mid O\}|\{O|\cdot\}\}, \{O, \{O \mid \cdot\} \mid \{O, \{O \mid \cdot\}|\cdot\}\} \ldots$$
$$\{O, \{\cdot \mid O\}|\{O|\cdot\}\} \ldots$$
$$\vdots$$

DAY t: The class $\mathfrak{R}_t$ of all those combinatorial games G with options in $\mathfrak{R}_{t-1}$ is created.

So one has $\mathfrak{R}_0 \subset \mathfrak{R}_1 \subset \ldots \subset \mathfrak{R}_t \subset \ldots$ and

$$\mathfrak{R} = \mathfrak{R}_0 \cup \mathfrak{R}_1 \cup \ldots \cup \mathfrak{R}_t \cup \ldots = \lim_{t \to \infty} \mathcal{R}_t.$$

Ex. 2.5. The number of combinatorial games grows rapidly:

(1) List all the combinatorial games in $\mathfrak{R}_2$.
(2) Argue that many more than 6000 combinatorial games exist at the end of DAY 3 (see Ex. 2.6).

Ex. 2.6. Show that $r_t = |\mathfrak{R}_t|$ grows super-exponentially fast:

$$r_t > 2^{r_{t-1}} \quad (t = 1, 2, \ldots)$$

(Hint: A finite set S with $n = |S|$ elements admits 2^n subsets.)

4. Winning strategies

A combinatorial game is started with either L or R making the first move. This determines the *first player*. The other player is the *second player*. The *normal* winning rule for an alternating 2-person games is:

> (NR) If a player $i \in \{L, R\}$ cannot move, player i has *lost* and the other player is declared the *winner*.

Chess matches, for example, are played under the normal rule: A loss of the king means a loss of the match (see Ex. 2.1).

REMARK 2.3 (Misère). The *misère* rule declares the player with no move to be the winner of the game.

A *winning strategy* for player i is a move (option) selection rule for i that ensures i to end as the winner.

> **THEOREM 2.1.** *In any combinatorial game G, an overall winning strategy exists for either the first or the second player.*

Proof. We prove the theorem by mathematical induction on $t = |G|$. In the case $t = 0$, we have

$$G = O = \{\cdot \mid \cdot\}.$$

Because the first player has no move in O, the second player is automatically the winner in normal play and hence has a winning strategy trivially guaranteed. Under the misère rule, the first player wins.

Suppose now $t \geq 1$ and that the theorem is true for all games that were created on DAY $t - 1$ or before. Consider the first player in G and assume that it is R. (The argument for L would go exactly the same way!)

If R has no option, L is the declared winner in normal play while R is the declared the winner in misère play. Either way, G has a guaranteed winner.

If options G^R exist, the induction hypothesis says that each of R's options leads to a situation in which either the first or the second player would have a winning strategy.

If there is (at least) one option G^R with the second player as the winner, R can take this option and win as the second player in G^R.

On the other hand, if all of R's options have their first player as the winner, there is nothing R can do to prevent L from winning. So the originally second player L has a guaranteed overall strategy to win the game. $\qquad\square$

Note that the proof of Theorem 2.1 is constructive in the following sense:

(1) Player i marks by $v(G^i) = +1$ all the options G^i in G that would have i winning as the then second player and sets $v(G^i) = -1$ otherwise.
(2) Player i follows the strategy to move to an option with the highest v-value.
(3) Provided a winning strategy exists at all for i, strategy (2) is a winning strategy for i.

The reader must be cautioned, however. The concrete computation of a winning strategy may be a very difficult task in real life.

Ex. 2.7 (DE BRUIJN's game). Two players choose a natural number $n \geq 1$ and write down all n numbers

$$1, 2, 3, \ldots, n-1, n.$$

A move of a player consists in selecting one of the numbers still present and erasing it together with all its (proper or improper) divisors.

Note that a winning strategy exists for the first player in normal play. Indeed, if it existed for the second player, the first player could simply erase "1" on the first move and afterwards (being now the second player) follow that strategy and win. Alas, no practically efficient method for the computation of a winning strategy is known.

REMARK 2.4. If chess is played with a finiteness rule, then a winning strategy exists for one of the two players. Currently, however, it is not known what it looks like. It is not even known which player is the potentially guaranteed winner.

Playing in practice. While winning strategies can be computed in principle (see the proof of Theorem 2.1), the combinatorial structure of many games is so complex that even today's computers cannot perform the computation efficiently.

In practice, a player i will proceed according to the following v-*greedy* strategy[1]:

> (vg_i) *Assign a quality estimate $v(G^i) \in \mathbb{R}$ to all the options G^i and move to an option with a highest v-value.*

A quality estimate v is not necessarily completely pre-defined by the game in absolute terms but may reflect previous experience and other considerations. Once quality measures are accepted as "reasonable", it is perhaps natural to expect that the game would evolve according to greedy strategies relative to these measures.

EX. 2.8. A popular rule of thumb evaluates the quality of a chess configuration σ for a player W, say, by assigning a numerical weight v

---

[1] Also chess computer programs follow this idea.

to the white pieces on the board. For example:

$$v(\text{pawn}) = 1$$
$$v(\text{bishop}) = 3$$
$$v(\text{knight}) = 3$$
$$v(\text{castle}) = 4.5$$
$$v(\text{queen}) = 9.$$

Where $v(\sigma)$ is the total weight of the white pieces, a v-greedy player W would choose a move to a configuration σ with a maximal value $v(\sigma)$. (Player B can, of course, evaluate the black pieces similarly.)

5. Algebra of games

For the rest of the chapter we will (unless explicitly said otherwise) assume:

- *The combinatorial games under consideration are played with the normal winning rule.*

The set $\mathfrak{R}$ of combinatorial games carries an algebraic structure which allows us to do computations with games as generalized numbers. This section wants to give a short sketch of the idea.[2]

Negation. We first define the *negation* $(-G)$ for the game

$$G = \{G_1^L, G_2^L, \ldots \mid G_1^R, G_2^R, \ldots\} \in \mathfrak{R}$$

as the game that arises from G when the players L and R interchange their roles: L becomes the "right" and R the "left" player.

---

[2](Much) more can be found in the highly recommended treatise of CONWAY [7].

So we obtain the negated games recursively as

$$-O = O$$
$$-G = \{-G_1^R, -G_2^R \dots \mid -G_1^L, -G_2^L, \dots\} \quad \text{if } G \neq O.$$

Also $-G$ is a combinatorial game and one has the straightforward algebraic rule

$$-(-G) = G.$$

Addition. The *sum* $G + H$ of the games G and H is the game in which a player $i \in \{L, R\}$ may choose to play either on G or on H. This means that i chooses an option G^i in G *or* an option H^i in H and accordingly reduces the game

$$\text{either to} \quad G^i + H \quad \text{or to} \quad G + H^i.$$

The reader is invited to verify the further algebraic rules:

$$G + H = H + G$$
$$(G + H) + K = G + (H + K)$$
$$G + O = G.$$

Moreover, we write

$$G - H = G + (-H).$$

Ex. 2.9. The second player wins $G - G \ (= G + (-G))$ in normal play with the obvious strategy:

- Imitate every move of the first player:
  When the first player chooses the option G^i in G, the second player will answer with the option $(-G^i)$ in $(-G)$, etc.

5.1. *Congruent games*

Motivated by Ex. 2.9, we say that combinatorial games G and H are *congruent* (notation: "$G \equiv H$") if

(C) $G - H$ can be won by the second player (in normal play).

In particular, $G \equiv O$ means that the second player has a winning strategy for G.

THEOREM 2.2 (Congruence Theorem). *For all $G, H, K \in \mathfrak{R}$, one has:*

(a) *If $G \equiv H$, then $H \equiv G$.*
(b) *If $G \equiv H$, then $G + K \equiv H + K$.*
(c) *If $G \equiv H$ and $H \equiv K$, then $G \equiv K$.*

Proof. The verification of the commutativity rule (a) is left to the reader. To see that (b) is true, we consider the game

$$M = (G + K) - (H + K)$$
$$= G + K - H - K$$
$$= (G - H) + (K - K).$$

The game $K - K$ can always be won by the second player (Ex. 2.9). Hence, if the second player can win $G - H$, then clearly M as well:

- *It suffices for the second player to apply the respective winning strategies to $G - H$ and to $K - K$.*

The proof of the transitivity rule (c) is similar. By assumption, the game

$$T = (G - K) + (-H + H) = (G - H) + (H - K)$$

can be won by the second player. We must show that the second player can therefore win $G - K$.

Suppose to the contrary that $G - K \not\equiv O$ were true and that the game $G - K$ could be won by the first player. Then the first player could win T by beginning with a winning move in $G - K$ and continuing with the win strategy whenever the second player moves in $G - K$. If the second player moves in $K - K$, the first player becomes second there and thus is assured to win on $K - K$! So the first player would win T, which would contradict the assumption however.

Hence we conclude that $G - K \equiv O$ must hold.   □

Congruence classes. For any $G \in \mathfrak{R}$, the class of congruent games is

$$[G] = \{H \in \mathfrak{R} \mid G \equiv H\}.$$

Theorem 2.2 says that addition and subtraction can be meaningfully defined for congruence classes:

$$[G] + [H] = [G + H] \quad \text{and} \quad [G] - [H] = [G - H].$$

In particular, we obtain the familiar algebraic rule

$$[G] - [G] = [G - G] = [O],$$

where $[O]$ is the class of all combinatorial games that are won by the second player. Hence we can re-cast the optimal strategy for a player (under the normality rule):

- *Winning strategy:*
  Make a move $G \to G'$ to an option $G' \in [O]$.

5.2. *Strategic equivalence*

Say that the combinatorial games G and H are *strategically equivalent* (denoted "$G \sim H$") if one of the following statements is true:

(SE_1) G and H can be won by the first player (*i.e.*, $G \not\equiv O \not\equiv H$).
(SE_2) G and H can be won by the second player (*i.e.*, $G \equiv O \equiv H$).

THEOREM 2.3 (Strategic equivalence). *Congruent games* $G, H \in \mathfrak{R}$ *are strategically equivalent,* i.e.,

$$G \equiv H \quad \Longrightarrow \quad G \sim H.$$

Proof. We claim that strategically non-equivalent games G and H cannot be congruent.

So assume, for example, that the first player wins G (*i.e.*, $G \not\equiv O$), and the second player wins H (*i.e.*, $H \equiv O$ and hence $(-H) \equiv O$). We will argue that the first player has a winning strategy for $G - H$, which means $G \not\equiv H$.

Indeed, the first player can begin with a winning strategy on G. Once the second player moves on $(-H)$, the first player, being now the second player on $(-H)$, wins there. Thus an overall victory is guaranteed for the first player. □

6. Impartial games

A combinatorial game G is said to be *impartial* (or *neutral*) if both players have the same options. The formal definition is recursive:

- $O = \{\cdot \mid \cdot\}$ is impartial.
- $G = \{A, B, \ldots, T \mid A, B, \ldots\}$ is impartial if all the options $A, B, \ldots, T$ are impartial.

Notice the following rules for impartial games G and H:

(1) $G = -G$ and hence $G + G = G - G \in [O]$.
(2) $G + H$ is impartial.

Ex. 2.10. Show that the frog game of Ex. 2.4 is impartial.

Nim is the prototypical impartial game (as the SPRAGUE–GRUNDY Theorem 2.4 will show below).

To formalize this claim, we use the notation $*n$ for a nim game relative to just one single set N_1 with $n = |N_1|$ elements. The options of $*n$ are the nim games

$$*0, *1, \ldots, *(n-1).$$

Moreover,

$$G = *n_1 + *n_2 + \cdots + *n_k$$

is the nim game described in Ex. 2.3 with k piles of sizes $n_1, n_2, \ldots, n_k$.

We now define the mex^3 of numbers $a, b, c \ldots, t$ as the smallest natural number g that equals none of the numbers $a, b, c, \ldots, t$:

$$\text{mex}\{a, b, c, \ldots, t\} = \min\{g \in \mathbb{N}_0 \mid g \notin \{a, b, c, \ldots, t\}\}, \qquad (12)$$

The crucial observation is stated in Lemma 2.1.

LEMMA 2.1. *For any finitely many numbers* $a, b, c, \ldots, t \in \mathbb{N}_0$, *one has*

$$G = \{*a, *b, *c, \ldots, *t \mid *a, *b, *c, \ldots, *t\} \equiv *\text{mex}\{a, b, c, \ldots, \},$$

i.e., the impartial game G *with the nim options* $*a, *b, *c,$ $\ldots, *t$ *is congruent with the simple nim game* $*m$ *with*

$$m = \text{mex}\{a, b, c, \ldots, t\}.$$

Proof. In view of $*m = -*m$, we must show: $G + *m \equiv O$, *i.e.*, the second player wins $G + *m$. Indeed, if the first player chooses an option $*j$ from

$$*m = \{*0, *1, \ldots, *(m-1)\},$$

then the second player can choose $*j$ from G (which must exist because of the definition of m as the minimal excluded number) and continue to win $*j + *j$ as the second player.

If the first player selects an option from G, say $*a$, we distinguish two cases. If $a > m$ then the second player reduces $*a$ to $*m$ and wins. If $a < m$, then the second player can reduce $*m$ to $*a$ and win. (Note that $a = m$ is impossible by the definition of mex.) $\qquad \square$

---

[3] "Minimal excluded".

> **THEOREM 2.4 (SPRAGUE–GRUNDY).** *Every impartial combinatorial game*
>
> $$G = \{A, B, C, \ldots \mid A, B, C \ldots, T\}$$
>
> *is congruent to a unique nim game of type $*m$. m is the so-called* GRUNDY *number of G and denoted by $\mathcal{G}(G)$.*
>
> $\mathcal{G}(G)$ *of can be computed recursively from the* GRUNDY *numbers of the options:*
>
> $$\mathcal{G}(G) = \operatorname{mex}\{\mathcal{G}(A), \mathcal{G}(B), \mathcal{G}(C), \ldots, \mathcal{G}(T)\}. \tag{13}$$

Proof. We prove the theorem by induction on $|G|$ and note $G \equiv O$ if $|G| = 0$. By induction, we now assume that the theorem is true for all options of G, *i.e.*, $A \equiv *a$, $B \equiv *b$ etc. with $a = \mathcal{G}(A)$, $b = \mathcal{G}(B)$, *etc.*

Hence we can argue $G \equiv *m = *\mathcal{G}(G)$ exactly as in the proof of Lemma 2.1. G cannot be congruent with another nim game $*k$ since (as Ex. 2.11 below shows):

$$*k \equiv *m \implies k = m. \qquad \square$$

Ex. 2.11. Show for all natural numbers k and n:

$$*k \equiv *n \iff k = n.$$

Ex. 2.12 (GRUNDY number of frogs). Let $F(n, k)$ be the (impartial) frog game of Ex. 2.4 and $\mathcal{G}(n, k)$ its GRUNDY number. For $k = 3$, $F(n, k)$ has the options

$$F(n - 1, 3), F(n - 2, 3), F(n - 3, 3).$$

So the associated GRUNDY number $\mathcal{G}(n, 3)$ has the recursion

$$\mathcal{G}(n, 3) = \operatorname{mex}\{\mathcal{G}(n - 1, 3), \mathcal{G}(n - 2, 3), \mathcal{G}(n - 3)\}.$$

Clearly, $\mathcal{G}(0,3) = 0, \mathcal{G}(1,3) = 1$ and $\mathcal{G}(2,3) = 2$. The recursion then produces the subsequent GRUNDY numbers:

n	0	1	2	3	4	5	6	7	8	9	10	$\cdots$
$\mathcal{G}(n,3)$	0	1	2	3	0	1	2	3	0	1	2	$\cdots$

The second player wins the nim game $*m$ if and only if $m = 0$. So the first player can win exactly the impartial games G with GRUNDY number $\mathcal{G}(G) \neq 0$. In general, we note:

- *Winning strategy for impartial games:*
  Make a move $G \to G'$ to an option G'
  with a GRUNDY number $\mathcal{G}(G') = 0$.

6.1. *Sums of* GRUNDY *numbers*

If G and H are impartial games with GRUNDY numbers $m = \mathcal{G}(G)$ and $n = \mathcal{G}(H)$, the GRUNDY number of their sum is

$$\mathcal{G}(G + H) = \mathcal{G}(*n + *m)$$

Indeed, if $G \equiv *m$ and $H \equiv *n$, then $G + H \equiv *m + *n$ must hold.[4] For the study of sums, we may therefore restrict ourselves to nim games. Moreover, the fundamental property

$$\mathcal{G}(G + G) = \mathcal{G}(*n + *n) = \mathcal{G}(O) = 0$$

suggests to study sums in the context of *binary algebra.*

Binary algebra. Recall that every natural number n has a unique binary representation in terms of powers of 2,

$$n = \sum_{j=0}^{\infty} \alpha_j 2^j,$$

---

[4]Recall Theorem 2.2!

with binary coefficients $\alpha_j \in \{0, 1\}$. We define binary addition of 0 and 1 (as in Section 1.2.3) according to the rules

$$0 \oplus 0 = 0 = 1 \oplus 1 \quad \text{and} \quad 0 \oplus 1 = 1 = 1 \oplus 0$$

and extend it to natural numbers:

$$\left(\sum_{j=0}^{\infty} \alpha_j 2^j \right) \oplus \left(\sum_{j=0}^{\infty} \beta_j 2^j \right) = \sum_{j=0}^{\infty} (\alpha_j \oplus \beta_j) 2^j.$$

REMARK 2.5. Notice that the binary representation of a natural number has only finitely many non-zero summands. Indeed, $\alpha_j = 0$ must hold for all $j > \log_2 n$ if

$$n = \sum_{j=0}^{\infty} \alpha_j 2^j \quad \text{with } \alpha_j \in \{0, 1\}.$$

Ex. 2.13. Show for the binary addition of natural numbers m, n, k:

$$n \oplus m = m \oplus n$$
$$n \oplus (m \oplus k) = (n \oplus m) \oplus k$$
$$n \oplus m \oplus k = 0 \iff n \oplus m = k.$$

The sum theorem. We consider nim games with three piles of n, m and k objects, *i.e.*, sums of three single nim games $*n$, $*m$, and $*k$.

LEMMA 2.2. *For all numbers $n, m, k \in \mathbb{N}_0$, one has:*

(1) *If $n \oplus m \oplus k \neq 0$, then the first player wins $*n + *m + *k$.*
(2) *If $n \oplus m \oplus k = 0$, then the second player wins $*n + *m + *k$.*

Proof. We prove the lemma by induction on $n + m + k$ and note that the statements (1) and (2) are obviously true in the case

$$n + m + k = 0.$$

By induction, we now assume that the lemma is true for all $n', m', k' \in \mathbb{N}_0$ such that

$$n' + m' + k' < n + m + k.$$

We must then show that the lemma holds for n, m, k with the binary representations

$$n = \sum_{j=0}^{\infty} \alpha_j 2^j, \ m = \sum_{j=0}^{\infty} \beta_j 2^j, \ k = \sum_{j=0}^{\infty} \gamma_j 2^j.$$

In the case (1) with $n \oplus m \oplus k \neq 0$, there must be at least one j such that

$$\alpha_j \oplus \beta_j \oplus \gamma_j = 1.$$

Let J be the largest such index j. Two of these coefficients $\alpha_J, \beta_J, \gamma_J$ must be equal and the third one must have value 1. So suppose $\alpha_J = \beta_J$ and $\gamma_J = 1$, for example, which implies

$$n \oplus m < k \quad \text{and} \quad n + m + (n \oplus m) < n + m + k.$$

Let $k' = n \oplus m$. We claim that the first player can win by reducing $*k$ to $*k'$. Indeed, the induction hypothesis says that the lemma is true for n, m, k'. Since

$$n \oplus m \oplus k' = n \oplus m \oplus n \oplus m = 0,$$

property (2) guarantees a winning strategy for the second player in the reduced nim game

$$*n + *m + *k'.$$

But the latter is the originally first player! So statement (1) is found to be true.

In case (2), when $n \oplus m \oplus k = 0$, the first player must make a move on one of the three piles. Let us say that $*n$ is reduced to $*n'$. Because $n = m \oplus k$, we have

$$n' \neq m \oplus k \quad \text{and therefore} \quad n' \oplus m \oplus k \neq 0.$$

Because the lemma is assumed to be true for n', m, k, statement (1) guarantees a winning strategy for the first player in the reduced game

$$*n' + *m * k,$$

which is the originally second player. $\qquad \square$

THEOREM 2.5 (Sums of impartial games). *For any imp-artial combinatorial games G and H, one has*

$$\mathcal{G}(G + H) = \mathcal{G}(G) \oplus \mathcal{G}(H).$$

Proof. Let $n = \mathcal{G}(G)$ and $m = \mathcal{G}(H)$ and $k = n \oplus m$. Then $n \oplus m + k = 0$ holds. So Lemma 2.2 says that the second player wins

$$n * + * m * (n \oplus m),$$

which yields

$$G + H \equiv *n + *m \equiv *(n \oplus m).$$

Consequently, $n \oplus m$ must be the GRUNDY number of $G + H$. $\quad\square$

We illustrate Theorem 2.5 with the nim game

$$G = *1 + *3 * + * 5 + *7$$

of four piles with $1, 3, 5$ and 7 objects respectively. The binary representations of the pile sizes are

$$1 = 1 \cdot 2^0$$
$$3 = 1 \cdot 2^0 + 1 \cdot 2^1$$
$$5 = 1 \cdot 2^0 + 0 \cdot 2^1 + 1 \cdot 2^2$$
$$7 = 1 \cdot 2^0 + 1 \cdot 2^1 + 1 \cdot 2^2.$$

So the GRUNDY number of G is

$$\mathcal{G}(*1 + *3 * + *5 + *7) = 1 \oplus 3 \oplus 5 \oplus 7 = 0.$$

Hence G can be won by the second player in normal play.

Ex. 2.14. Suppose that the first player removes 3 objects from the pile of size 7 in $G = *1 + *3 * + *5 + *7$. How should the second player respond?

Ex. 2.15. There is a pile of 10 red and another pile of 10 black pebbles. Two players move alternatingly with the following options:

- EITHER: take at least 1 but not more than 3 of the red pebbles
  OR: take at least 1 but not more than 2 of the black pebbles.

Which of the players has a winning strategy in normal play? (Hint: Compute the GRUNDY numbers for the red and black piles separately (as in Ex. 2.4) and apply Theorem 2.5.)

Chapter 3

Zero-Sum Games

> Zero-sum games abstract the model of combinatorial games. Fundamental examples arise naturally as LAGRANGE games from mathematical optimization problems and thus furnish an important link between game theory and mathematical optimization theory. In particular, strategic equilibria in such games correspond to optimal solutions of optimization problems. Conversely, mathematical optimization techniques are important tools for the analysis of game-theoretic situations.

As in the previous chapter, we consider games Γ with 2 agents (or players). However, we shift the viewpoint and no longer assume players taking turns in moving a system from one state into another. Instead, we assume that the players decide on strategies according to which they play the game.

More precisely, the players are assumed to have sets X and Y of possible strategies (or decisions, or actions, *etc.*) at their disposal. Hence a "state" of the game under this new perspective is a pair (x, y) of strategies $x \in X$ and $y \in Y$.[1] In fact, we consider the set

$$\mathfrak{S} = \{(x, y) \mid x \in X, y \in Y\} \cup \{\sigma_0\}$$

of joint strategy choices as the system underlying the game Γ, where σ_0 is just an initial state, which we may attach formally to the set $X \times Y$ of joint strategies. We therefore refer to the one player as the *x-player* and to the other as the *y-player*. $\qquad$ •

---

[1] As in the Prisoner's dilemma (Ex. 1.10).

Γ is a *zero-sum game* if there is a function

$$U : X \times Y \to \mathbb{R}$$

that encodes the *utility* of the strategic choice (x, y) in the sense that the utility values of the individual players add up to zero:

(1) $u_1(x, y) = U(x, y)$ is the gain of the x-player;
(2) $u_2(x, y) = -U(x, y)$ is the gain of the y-player.

It follows that the two players have opposing goals:

(X) *The x-player wants to choose $x \in X$ as to maximize $U(x, y)$.*
(Y) *The y-player wants to choose $y \in Y$ as to minimize $U(x, y)$.*

We denote the corresponding zero-sum game by $\Gamma = \Gamma(X, Y, U)$.

Ex. 3.1. A combinatorial game with respective strategy sets X and Y for the two players is, in principle, a zero-sum game with the utility function

$$U(x, y) = \begin{cases} +1 & \text{if } x \text{ is a winning strategy for the } X\text{-player} \\ -1 & \text{if } y \text{ is a winning strategy for the } Y\text{-player} \\ 0 & \text{otherwise.} \end{cases}$$

Ex. 3.2. The Prisoner's dilemma is a 2-person game. However, the utilities of Ex. 1.10 do not lead to a zero-sum game.

1. Matrix games

In the case of finite strategy sets, say $X = \{1, \ldots, m\}$ and $Y = \{1, \ldots, n\}$, a function $U : X \times Y \to \mathbb{R}$ can be presented in matrix form:

$$U = \begin{bmatrix} u_{11} & u_{21} & \cdots & u_{1n} \\ u_{21} & u_{22} & \cdots & u_{2n} \\ \vdots & \vdots & & \vdots \\ u_{m1} & u_{m2} & \cdots & u_{mn} \end{bmatrix} \in \mathbb{R}^{m \times n}.$$

The associated *matrix game* is the zero-sum game $\Gamma = (X, Y, U)$ where the x-player chooses a row i and the y-player a column j.

This joint selection (i, j) has the utility value u_{ij} for the row player
and the value $(-u_{ij})$ for the column player. As an example, consider
the game with the utility matrix

$$U = \begin{bmatrix} +1 & -2 \\ -1 & +2 \end{bmatrix}. \tag{14}$$

In the example (14), there is no obvious overall "optimal" choice
of strategies. No matter what row i and column j are selected by
the players, one of the players will find[2] that the other choice would
have been more profitable. In this sense, this matrix game has no
"solution".

Before pursuing this point further, we will introduce a general
concept for the *solution* of a zero-sum game in terms of an equilibrium
between both players.

2. Equilibria

Let us assume that both players in the zero-sum game $\Gamma = (X, Y, U)$
are risk avoiding and want to ensure themselves optimally against
the worst case. So they consider the worst case functions

$$\begin{aligned} U_1(x) &= \min_{y \in Y} u(x, y) \in \mathbb{R} \cup \{-\infty\} \\ U_2(y) &= \max_{x \in X} u(x, y) \in \mathbb{R} \cup \{+\infty\}. \end{aligned} \tag{15}$$

The x-player thus faces the *primal problem*

$$\max_{x \in X} U_1(x) = \max_{x \in X} \min_{y \in Y} U(x, y), \tag{16}$$

while the y-player is to solve the *dual problem*

$$\min_{y \in Y} U_2(y) = \min_{y \in Y} \max_{x \in X} U(x, y). \tag{17}$$

---

[2]In hindsight!

From the definition, one immediately deduces for any $x \in X$ and $y \in Y$ the *primal-dual inequality*:

$$U_1(x) \leq U(x, y) \leq U_2(y) \tag{18}$$

We say that $(x^*, y^*) \in X \times Y$ is an *equilibrium* of the game Γ if it yields the equality

$$U_1(x^*) = U_2(y^*),$$

i.e., if the primal-dual inequality, in fact, attains equality:

$$\max_{x \in X} \min_{y \in Y} \; U(x, y) = U(x^*, y^*) = \min_{y \in Y} \max_{x \in X} \; U(x, y) \tag{19}$$

In the equilibrium (x^*, y^*), none of the risk avoiding players has an incentive[3] to deviate from the chosen strategy. In this sense, equilibria represent *optimal* strategies for risk avoiding players.

Ex. 3.3. Determine the best worst-case strategies for the two players in the matrix game with the matrix U of (14) and show that the game has no equilibrium.

Give furthermore an example of a matrix game that possesses at least one equilibrium.

Finding equilibria. If the strategy sets X and Y are finite and hence the zero-sum game $\Gamma = (X, Y, U)$ is a matrix game, the question whether an equilibrium exists, can — in principle — be answered in finite time by a simple procedure:

- *Check each strategy pair $(x^*, y^*) \in X \times Y$ for the property (19).*

---

[3]In hindsight.

If X and Y are infinite, usually the existence of equilibria can only be decided if the function $U : X \times Y \to \mathbb{R}$ has special properties. From a theoretical point of view, the notion of *convexity* is very helpful and important.

3. Convex zero-sum games

Recall[4] that a *convex combination* of points $x_1, \ldots, x_k \in \mathbb{R}^n$ is a linear combination

$$\overline{x} = \sum_{i=1}^{k} \lambda_i x_i \quad \text{with coefficients } \lambda_i \geq 0 \text{ such that } \sum_{i=1}^{k} \lambda_i = 1.$$

An important interpretation of $\overline{x}$ is based on the observation that the coefficient vector $\lambda = (\lambda_1, \ldots, \lambda_k)$ of a convex combination is a probability distribution:

- If a point x_i is selected from the set $\{x_1, \ldots, x_k\}$ with probability λ_i, then the components of the convex combination $\overline{x}$ are exactly the expected component values of the stochastically selected point.

Another way of looking at $\overline{x}$ is:

- If weights of size λ_i are placed on the points x_i, then $\overline{x}$ is their center of gravity.

A set $X \subseteq \mathbb{R}^n$ is *convex* if X contains all convex combinations of all possible finite subsets $\{x_1, \ldots, x_k\} \subseteq X$.

Ex. 3.4. Let $S = \{s_1, \ldots, s_m\}$ be an arbitrary set with $m \geq 1$ elements. Show that the set $\overline{S}$ of all probability distributions λ on S forms a compact convex subset of $\mathbb{R}^m$.

---

[4]See also Appendix A.2 for more details.

A function $f : X \to \mathbb{R}$ is *convex* (or *convex up*) if X is a convex subset of some coordinate space $\mathbb{R}^n$ and for every $x_1, \ldots, x_k \in X$ and probability distribution $\lambda = (\lambda_1, \ldots, \lambda_k)$, one has

$$f(\lambda_1 x_1 + \ldots + \lambda_k x_k) \le \lambda_1 f(x_1) + \ldots + \lambda_k f(x_k).$$

f is *concave* (or *convex down*) if $g = -f$ is convex (up).

With this terminology, we say that the zero-sum game $\Gamma = (X, Y, U)$ is *convex* if

---

(1) X and Y are non-empty convex strategy sets;
(2) the utility $U : X \times Y \to \mathbb{R}$ is such that

   (a) for every $y \in Y$, the map $x \mapsto U(x, y)$ is concave.
   (b) for every $x \in X$, the map $y \mapsto U(x, y)$ is convex.

The main theorem on general convex zero-sum-games guarantees the existence of at least one equilibrium in the case of compact strategy sets:

THEOREM 3.1. *A convex zero-sum game* $\Gamma = (X, Y, U)$ *with compact strategy sets* X *and* Y *and a continuous utility* U *admits a strategic equilibrium* $(x^*, y^*) \in X \times Y$.

Proof. Since X and Y are convex and compact sets, so is the set $Z = X \times Y$ and hence also the set $Z \times Z$.

Consider the continuous function $G : Z \times Z \to \mathbb{R}$ with the values

$$G((x', y'), (x, y)) = U(x, y') - U(x', y).$$

Since U is concave in the first variable x and $(-U)$ concave in the second variable y, we find that G is concave in the second variable (x, y). Hence Corollary A.1 (Appendix) allows us to deduce the existence of an element $(x^*, y^*) \in Z$ that satisfies

$$\begin{aligned} 0 &= G((x^*, y^*), (x^*, y^*)) \\ &\ge G((x^*, y^*), (x, y)) = U(x, y^*) - U(x^*, y) \end{aligned}$$

for all $(x, y) \in Z$ and hence the inequality

$$U(x, y^*) \leq U(x^*, y) \quad \text{for all } x \in X \text{ and all } y \in Y.$$

This shows that x^* is the best strategy for the x-player if the y-player chooses $y^* \in Y$. Similarly, y^* can be proved to be optimal against x^*. In other words, (x^*, y^*) is an equilibrium of (X, Y, U). $\qquad \square$

Theorem 3.1 has important consequences not only in game theory but also in the theory of mathematical optimization in general, which we will sketch in more details in Section 4 below. To illustrate the situation, let us first look at the special case of randomizing the strategic decisions in zero-sum matrix games.

3.1. *Randomized matrix games*

Recall from Ex. 5.2 that a zero-sum game $\Gamma = (X, Y, U)$ with finite strategy sets X and Y does not necessarily admit an equilibrium.

Suppose the players *randomize* the choice of their respective strategies. That is to say, the x-player decides on a probability distribution $\overline{x}$ on X and chooses an $i \in X$ with probability $\overline{x}_i$. Similarly, the y-player chooses a probability distribution $\overline{y}$ on Y and selects $j \in Y$ with probability $\overline{y}_j$. Then the x-player's expected gain is

$$\overline{U}(\overline{x}, \overline{y}) = \sum_{i \in X} \sum_{j \in Y} u_{ij} \overline{x}_i \overline{y}_j,$$

where the u_{ij} are the coefficients of the utility matrix U.

So we arrive at a zero-sum game $\overline{\Gamma} = (\overline{X}, \overline{Y}, \overline{U})$, where $\overline{X}$ is the set of probability distributions on X and $\overline{Y}$ the set of probability distributions on Y. $\overline{X}$ and $\overline{Y}$ are compact convex sets (*cf.* Ex. 3.4).

Moreover, the function $\overline{U} : \overline{X} \times \overline{Y} \to \mathbb{R}$ is linear, and thus continuous and both concave and convex in both components.

It follows that $\overline{\Gamma}$ is a convex game that satisfies the hypothesis of Theorem 3.1. Therefore, $\overline{\Gamma}$ admits an equilibrium. This proves the VON NEUMANN's Theorem:

THEOREM 3.2 (VON NEUMANN [32]). *Let $U \in \mathbb{R}^{m \times n}$ be an arbitrary matrix with coefficients u_{ij}. Then there exist $x^* \in X$ and $y^* \in Y$ such that*

$$\max_{1 \leq i \leq m} \min_{1 \leq y \leq m} \sum_{i,j} u_{ij} x_i y_j = \sum_{i,j} u_{ij} x_i^* y_j^* = \min_{1 \leq j \leq n} \max_{1 \leq i \leq m} \sum_{i,j} u_{ij} x_i y_j.$$

where X is the set of all probability distributions on $\{1, \ldots, m\}$ and Y the set of all probability distributions on $\{1, \ldots, n\}$.

3.2. *Computational aspects*

While it is generally not easy to compute equilibria in zero-sum games, the task becomes tractable for randomized matrix games. Consider, for example, the two sets $X = \{1, \ldots, m\}$ and $Y = \{1, \ldots, n\}$ and the utility matrix

$$U = \begin{bmatrix} u_{11} & u_{21} & \cdots & u_{1n} \\ u_{21} & u_{22} & \cdots & u_{2n} \\ \vdots & \vdots & & \vdots \\ u_{m1} & u_{m2} & \cdots & u_{mn} \end{bmatrix} \in \mathbb{R}^{m \times n}$$

For the probability distributions $x \in \overline{X}$ and $y \in \overline{Y}$, the expected utility for the x-player is

$$\overline{U}(x,y) = \sum_{i=1}^{m} \sum_{j=1}^{n} u_{ij} x_i y_j = \sum_{j=1}^{n} y_j \left(\sum_{i=1}^{m} u_{ij} x_i \right).$$

The worst case for the row player occurs when the column player selects a probability distribution that puts the full weight 1 on $k \in Y$ such that

$$\sum_{i=1}^{m} u_{ik} x_i = \min \left\{ \sum_{i=1}^{m} u_{ij} x_i \mid j = 1, \ldots, n \right\} = \overline{U}_1(x).$$

Hence

$$\max_{x \in \overline{X}} U_1(x) = \max_{z \in \mathbb{R}, x \in \overline{X}} \left\{ z \mid z \le \sum_{i=1}^{m} u_{ij} x_i \text{ for all } i = 1, \dots, m \right\}. \tag{20}$$

Similarly, the worst case for the column player it attained when the row player places the full probability weight 1 onto $\ell \in X$ such that

$$\sum_{j=1}^{n} u_{\ell j} y_j = \max \left\{ \sum_{j=1}^{n} u_{ij} y_j \mid i = 1, \dots, m \right\} = \overline{U}_2(y).$$

This yields

$$\min_{y \in \overline{Y}} \overline{U}_2(y) = \min_{w \in \mathbb{R}, y \in \overline{Y}} \left\{ w \mid w \ge \sum_{j=1}^{n} u_{ij} y_j \text{ for all } j = 1, \dots, n \right\}. \tag{21}$$

This analysis shows:

PROPOSITION 3.1. *If (z^*, x^*) is an optimal solution of (20) and (w^*, y^*) and optimal solution of (21), then*

(1) (x^*, y^*) *is an equilibrium of* $\overline{\Gamma} = (\overline{X}, \overline{Y}, \overline{U})$.
(2) $z^* = \max_{x \in \overline{X}} U_1(x) = \min_{y \in \overline{Y}} U_2(y) = w^*$.

REMARK 3.1. As further outlined in Section 4.5 below, the optimization problems (20) and (21) are *linear programs* that are dual to each other. They can be solved very efficiently in practice. For explicit solution algorithms, the interested reader may consult the standard literature on mathematical optimization.[5]

---

[5] *e.g.*, FAIGLE *et al.* [17].

4. LAGRANGE games

The analysis of zero-sum games is closely connected with a fundamental technique in mathematical optimization. A very general formulation of an optimization problem is

$$\max_{x \in \mathcal{F}} f(x),$$

where $\mathcal{F}$ could be any set and $f : \mathcal{F} \to \mathbb{R}$ an arbitrary *objective function*. In our context, however, we will look at more concretely specified problems and understand by a *mathematical optimization problem* a problem of the form

$$\max_{x \in X} f(x) \quad \text{such that} \quad g(x) \geq 0, \tag{22}$$

where X is a nonempty subset of some coordinate space $\mathbb{R}^n$ with an objective function $f : X \to \mathbb{R}$.

The vector valued function $g : X \to \mathbb{R}^m$ is a *restriction* function and combines m real-valued restriction functions $g_i : X \to \mathbb{R}$ as its components. The set of *feasible solutions* of (22) is

$$\mathcal{F} = \{x \in X \mid g_i(x) \geq 0 \text{ for all } i = 1, \ldots, m\}.$$

REMARK 3.2. The model (22) formulates an optimization problem as a maximization problem. Minimization problems can, of course, also be formulated within this model because of

$$\min_{x \in \mathcal{F}} f(x) = -\max_{x \in \mathcal{F}} \tilde{f}(x) \quad \text{with the objective } \tilde{f}(x) = -f(x).$$

The optimization problem (22) defines a zero-sum game $\Lambda = (X, \mathbb{R}^m_+, L)$ with the so-called LAGRANGE *function*

$$L(x, y) = f(x) + y^T g(x) = f(x) + \sum_{i=1}^{m} y_i g_i(x) \tag{23}$$

as its utility. We refer to Λ as a LAGRANGE *game*.[6] The worst-case utility functions of the two players in Λ are:

$$L_1(x) = \min_{y \geq 0} \ L(x,y) = f(x) + \min_{y \geq 0} \ \sum_{i=1}^{m} y_i g_i(x)$$

$$L_2(y) = \max_{x \in X} \ L(x,y) = \max_{x \in X} \ f(x) + \sum_{i=1}^{m} y_i g_i(x).$$

Ex. 3.5 (Convex LAGRANGE games). If X is convex and the objective function $f : X \to \mathbb{R}$ as well as the restriction functions $g_i : X \to \mathbb{R}$ are concave, then the LAGRANGE game $\Lambda = (X, \mathbb{R}_+^m, L)$ is a convex zero-sum game. Indeed, $L(x,y)$ is concave in x for every fixed $y \geq 0$ and linear in y for every fixed $x \in X$. Since linear functions are in particular convex, the game Λ is convex.

4.1. *Complementary slackness*

The choice of an element $x \in X$ with at least one restriction violation $g_i(x) < 0$ would allow the y-player in the LAGRANGE game $\Lambda = (X, \mathbb{R}_+^m, L)$ to increase its utility value infinitely with the choice $y_i \approx \infty$. So the risk avoiding x-player will always try to select a feasible x.

On the other hand, if all of the feasibility requirements $g_i(x) \geq 0$ are satisfied, the best the y-player can do is the selection of $y \in \mathbb{R}_+^m$ such that the so-called *complementary slackness condition*

$$\sum_{i=1}^{m} y_i g_i(x) = y^T g(x) = 0 \quad \text{and hence} \quad L(x,y) = f(x) \quad (24)$$

is met. Consequently, one finds:

---

[6]The idea goes back to J.-L. LAGRANGE (1736–1813).

The primal LAGRANGE problem is identical with the original problem:

$$\max_{x \in X} \min_{y \geq 0} L(x, y) = \max_{x \in \mathcal{F}} L_1(x) = \max_{x \in \mathcal{F}} f(x). \qquad (25)$$

LEMMA 3.1. If (x^*, y^*) is an equilibrium of the LAGRANGE game Λ, then x^* is an optimal solution of problem (22).

Proof. For every feasible $x \in \mathcal{F}$, $y^* \geq 0$ yields $(y^*)^T g(x) \geq 0$ and, therefore,

$$f(x^*) = L_1(x^*) = L_2(y^*) \geq f(x) + (y^*)^T g(x) \geq f(x).$$

So x^* is optimal. $\square$

4.2. The KKT-conditions

Lemma 3.1 indicates the importance of being able to identify equilibria in LAGRANGE games. In order to establish necessary conditions, *i.e.*, conditions which candidates for equilibria must satisfy, we impose further assumptions on problem (22):

(1) $X \subseteq \mathbb{R}^n$ is a convex set, *i.e.*, X contains with every x, x' also the whole line segment

$$[x, x'] = \{x + \lambda(x' - x) \mid 0 \leq \lambda \leq 1\}.$$

(2) The functions f and g_i in (22) have continuous partial derivatives $\partial f(x)/\partial x_j$ and $\partial g_i(x)/\partial x_j$ for all $j = 1, \ldots, n$.

It follows that also the partial derivatives of the LAGRANGE function L exist. So the marginal change of L into the direction d of the x-variables is

$$\nabla_x L(x, y)d = \nabla f(x)d + \sum_{i=1}^{m} y_i \nabla g_i(x)d$$

$$= \sum_{j=1}^{n} \frac{\partial f(x)}{\partial x_j} d_j + \sum_{i=1}^{m} \sum_{j=1}^{n} \frac{\partial g_i(x)}{\partial x_j} y_i d_j.$$

REMARK 3.3 (JACOBI matrix). The $(m \times n)$ matrix $Dg(x)$ having as coefficients the partial derivatives

$$Dg(x)_{ij} = \partial g_i(x)/\partial x_j$$

of a function $g : \mathbb{R}^n \to \mathbb{R}^m$ is known as a *functional* or JACOBI[7] *matrix*. It provides a compact matrix notation for the marginal change of the *Lagrange* function:

$$\nabla_x L(x, y)d = \nabla f(x) + y^T Dg(x)d.$$

LEMMA 3.2 (KKT-conditions). *The pair* $(x, y) \in X \times \mathbb{R}_+^m$ *cannot be an equilibrium of the* LAGRANGE *game* Λ *unless:*

(K_0) $g(x) \geq 0$;
(K_1) $y^T g(x) = 0$;
(K_3) $\nabla_x L(x, y)d \leq 0$ *holds for all* d *such that* $x + d \in X$.

Proof. We already know that the feasibility condition (K_0) and the complementary slackness condition (K_1) are necessarily satisfied by an equilibrium. If (K_3) were violated and $\nabla_x L(x, y)d > 0$ were true, the x-player could improve the L-value by moving a bit into direction d. This would contradict the definition of an "equilibrium". $\qquad \square$

REMARK 3.4. The three conditions of Lemma 3.2 are the so-called *KKT-conditions*.[8] Although they are always necessary, they are not

---

[7]C.G. JACOBI (1804–1851).
[8]Named after the mathematicians KARUSH, KUHN and TUCKER.

always sufficient to conclude that a candidate (x, y) is indeed an equilibrium.

4.3. Shadow prices

Given m functions $a_1, \ldots, a_m : \mathbb{R}_+^n \to \mathbb{R}$ and m scalars $b_1, \ldots,$
$b_m \in \mathbb{R}$, the optimization problem

$$\max_{x \in \mathbb{R}_+^n} f(x) \quad \text{s.t.} \quad a_1(x) \le b_1, \ldots, a_m(x) \le b_m \tag{26}$$

is of type (22) with the m restriction functions $g_i(x) = b_i - a_i(x)$ and has the LAGRANGE function

$$
\begin{aligned}
L(x, y) &= f(x) + \sum_{i=1}^{m} y_i(b_i - a_i(x)) \\
&= f(x) - \sum_{i=1}^{m} y_i a_i(x) + \sum_{i=1}^{m} y_i b_i.
\end{aligned}
$$

For an intuitive interpretation of the problem (26), think of the data vector

$$x = (x_1, \ldots, x_n)$$

as a plan for n products to be manufactured in quantities x_j and of $f(x)$ as the market value of x.

Assume that x requires the use of m materials in quantities $a_1(x), \ldots, a_m(x)$ and that the parameters $b_1, \ldots, b_m$ describe the quantities of the materials already in the possession of the manufacturer.

If the numbers $y_1, \ldots, y_m$ represent the market prices (per unit) of the m materials, we find that $L(x, y)$ is the total value of the manufacturer's assets:

$$
\begin{aligned}
L(x, y) = {} & \text{market value of the production } x \\
& + \text{value of the materials left in stock.}
\end{aligned}
$$

The manufacturer would like to have that value as high as possible by deciding on an appropriate production plan x.

"The market" is an opponent of the manufacturer and looks at the value

$$-L(x, y) = \sum_{i=1}^{m} y_i(a_i(x) - b_i) - f(x),$$

which is the value of the materials the manufacturer must still buy on the market for the production of x minus the value of the production that the market would have to pay to the manufacturer for the production x.

The market would like to set the prices y_i so that $-L(x, y)$ is as large as possible. Hence:

- *The manufacturer and the market play a* LAGRANGE *game* Λ.
- An equilibrium (x^*, y^*) of Λ reflects an economic balance: *Neither the manufacturer nor the market has a guaranteed way to improve their value by changing the production plan or by setting different prices.*

In this sense, the production plan x^* is *optimal* if (x^*, y^*) is an equilibrium. The optimal market prices $y_1^*, \ldots, y_m^*$ are the so-called *shadow prices* of the m materials.

The complementary slackness condition (K_1) says that a material which is in stock but not completely used by x^* has zero market value:

$$a_i(x^*) < b_i \quad \Longrightarrow \quad y_i^* = 0.$$

The condition (K_2) implies that x^* is a production plan of optimal value $f(x^*) = L(x^*, y^*)$ under the given restrictions. Moreover, one has

$$\sum_{i=1}^{m} y_i^* a_i(x^*) = \sum_{i=1}^{m} y_i^* b_i,$$

which says that the price of the materials used for the production x^* equals the value of the inventory under the shadow prices y_i^*.

Property (K₃) says that the marginal change $\nabla_x L(x^*, y^*)d$ of the manufacturer's value L is negative in any feasible production modification from x^* to $x^* + d$ and only profitable for the market because

$$\nabla_x(-L(x^*, y^*)) = -\nabla_x L(x^*, y^*).$$

We will return to production games in the context of cooperative game theory in Section 8.3.2.

4.4. Equilibria of convex LAGRANGE games

Remarkably, the KKT-conditions turn out to be not only necessary but also sufficient for the characterization of equilibria in convex LAGRANGE games with differentiable objective functions. This gives a way to compute such equilibria and hence to solve optimization problems of type (22) in practice[9]:

- *Find a solution* $(x^*, y^*) \in X \times \mathbb{R}^m_+$ *for the KKT-inequalities.* (x^*, y^*) *will yield an equilibrium in* $\Lambda = (X, \mathbb{R}^m_+, L)$ *and* x^* *will be an optimal solution for* (22).

Indeed, one finds:

THEOREM 3.3. *A pair* $(x^*, y^*) \in X \times \mathbb{R}^m_+$ *is an equilibrium of the convex LAGRANGE game* $\Lambda = (X, \mathbb{R}^m_+, L)$ *if and only if* (x^*, y^*) *satisfies the KKT-conditions.*

Proof. From Lemma 3.2, we know that the KKT-conditions are necessary. To show sufficiency, assume that $(x^*, y^*) \in X \times \mathbb{R}^m_+$ satisfies the KKT-conditions. We must demonstrate that (x^*, y^*) is

---

[9]It is not the current purpose to investigate further computational aspects in details, which can be found in the established literature on mathematical programming.

an equilibrium of the convex LAGRANGE game $\Lambda = (X, \mathbb{R}^m_+, L)$, *i.e.*, satisfies

$$\max_{x \in X} L(x, y^*) = L(x^*, y^*) = \min_{y \geq 0} L(x^*, y) \qquad (27)$$

for the function $L(x, y) = f(x) + y^T g(x)$.

Since $x \mapsto L(x, y^*)$ is a concave function, we have for every $x \in X$,

$$L(x, y^*) - L(x^*, y^*) \leq \nabla_x L(x^*, y^*)(x - x^*).$$

Because (K_2) guarantees $\nabla_x L(x^*, y^*)(x - x^*) \leq 0$, we conclude

$$L(x, y^*) \leq L(x^*, y^*),$$

which implies the first equality in (27). For the second equality, (K_0) and (K_1) yield $g(x^*) \geq 0$ and $(y^*)^T g(x^*) = 0$ and therefore:

$$\min_{y \geq 0} L(x^*, y) = f(x^*) + \min_{y \geq 0} y^T g(x^*) = f(x^*) + 0$$
$$= f(x^*) + (y^*)^T g(x^*) = L(x^*, y^*). \qquad \square$$

4.5. *Linear programs*

A *linear program* (LP) in *standard form* is an optimization problem of the form

$$\max_{x \in \mathbb{R}^n_+} c^T x \quad \text{s.t.} \quad Ax \leq b, \qquad (28)$$

where $c \in \mathbb{R}^n$ and $b \in \mathbb{R}^m$ are parameter vectors and $A \in \mathbb{R}^{m \times n}$ a matrix, and thus is a mathematical optimization problem with a linear objective function $f(x) = c^T x$ and restriction function $g(x) = b - Ax$ so that

$$g(x) \geq 0 \quad \longleftrightarrow \quad Ax \leq b.$$

The feasibility region $\mathcal{F}$ of (28) is the set of all nonnegative solutions of the linear inequality system $Ax \leq b$:

$$\mathcal{F} = P_+(A, b) = \{x \in \mathbb{R}^n \mid Ax \leq b, x \geq 0\}.$$

The LAGRANGE function is

$$L(x, y) = c^T x + y^T (b - Ax) = y^T b + (c^T - y^T A)x$$

and yields for any $x \geq 0$ and $y \geq 0$:

$$L_1(x) = \min_{y \geq 0} L(x, y) = \begin{cases} c^T x & \text{if } Ax \leq b \\ -\infty & \text{if } Ax \not\leq b. \end{cases}$$

$$L_2(y) = \max_{x \geq 0} L(x, y) = \begin{cases} b^T y & \text{if } y^T A \geq c^T \\ +\infty & \text{if } y^T A \not\geq c^T. \end{cases}$$

The optimum value of L_2 is found by solving the *dual* associated linear program

$$\min_{y \geq 0} L_2(y) = \min_{y \geq 0} y^T b \quad \text{s.t.} \quad y^T A \geq c^T. \tag{29}$$

Ex. 3.6. Since $y^T b = b^T y$ and $(c^T - y^T A)x = x^T(c - A^T y)$ holds, the dual linear program (29) can be formulated equivalently in standard form:

$$\max_{y \in \mathbb{R}_+^m} (-b)^T y \quad \text{s.t.} \quad (-A^T)y \leq -c. \tag{30}$$

The main theorem on linear programming is:

THOREM 3.4 (Main LP-Theorem). *For the LP (28) the following holds:*

(A) *An optimal solution x^* exists if and only if both the LP (28) and the dual LP (29) have feasible solutions.*

(B) *A feasible x^* is an optimal solution if and only if there exists a dually feasible solution y^* such that*

$$c^T x^* = L_1(x^*) = L_2(y^*) = b^T y^*.$$

Proof. Assume that (28) has an optimal solution x^* with value $z^* = c^T x^*$. Then $c^T x \leq z^*$ holds for all feasible solutions x. So the

FARKAS Lemma[10] guarantees the existence of some $y^* \geq 0$ such that

$$(y^*)^T A \geq c^T \quad \text{and} \quad (y^*)^T b \leq z^*.$$

Noticing that y^* is dually feasible and that $L_1(x^*) \leq L_2(y)$ holds for all $y \geq 0$, we conclude that y^* is, in fact, an optimal dual solution:

$$L_2(y^*) = (y^*)^T b \leq z^* = L_1(x^*) \leq L_2(y^*) \quad \Longrightarrow \quad L_1(x^*) = L_2(y^*).$$

This argument establishes property (B) and shows that the existence of an optimal solution necessitates the existence of a dually feasible solution. Assuming that (28) has at least one feasible solution x, it therefore remains to show that the existence a dual feasible solution y implies the existence of an optimal solution.

To see this, note first

$$w^* = \inf_{y \geq 0} b^T y \geq L_1(x) > -\infty.$$

So each dually feasible y satisfies $-b^T y \leq -w^*$. Applying now the FARKAS Lemma to the dual linear program in the form (30), we find that a parameter vector $x^* \geq 0$ exists with the property

$$Ax^* \leq b \quad \text{and} \quad L_1(x^*) \geq w^*.$$

On the other hand, the primal-dual inequality yields $L_1(x^*) \leq w^*$. So x^* must be an optimal feasible solution. □

General linear programs. In general, a *linear program* refers to the problem of optimizing a linear objective function over a *polyhedron*, namely the set of solutions of a finite system of linear equalities and inequalities and, therefore, can be formulated as

$$\max_{x \in \mathbb{R}^n} c^T x \quad \text{s.t.} \quad Ax \leq b, Bx = d \tag{31}$$

with coefficient vectors $c \in \mathbb{R}^n$, $b \in \mathbb{R}^m$, $d \in \mathbb{R}^k$ and matrices $A \in \mathbb{R}^{m \times n}$ and $B \in \mathbb{R}^{k \times n}$.

---

[10]See Lemma A.6 in the Appendix.

If no equalities occur in the formulation (31), one has as a linear program in *canonical form*:

$$\max_{x \in \mathbb{R}^n} c^T x \quad \text{s.t.} \quad Ax \leq b. \tag{32}$$

Because of the equivalence

$$Bx = d \quad \Longleftrightarrow \quad Bx \leq d \text{ and } -B \leq -d,$$

the optimization problem (31) can be presented in canonical form:

$$\max_{x \in \mathbb{R}^n} c^T x \quad \text{s.t.} \quad \begin{bmatrix} A \\ B \\ -B \end{bmatrix} x \leq \begin{pmatrix} b \\ d \\ -d \end{pmatrix}.$$

Moreover, since any vector $x \in \mathbb{R}^n$ can be expressed as the difference

$$x = x^+ - x^-$$

of two (nonnegative) vectors $x^+, x^- \in \mathbb{R}^n_+$, one sees that each linear program in canonical form is equivalent to a linear program in standard form:

$$\max_{x^+, x^- \geq 0} c^T x^- - c^T x^- \quad \text{s.t.} \quad Ax^+ - Ax^- \leq b.$$

The LAGRANGE function of the canonical form is the same as for the standard form. Since the domain of L_1 is now $X = \mathbb{R}^n$, the utility function $L_2(y)$ differs accordingly:

$$L_1(x) = \min_{y \geq 0} L(x, y) = \begin{cases} c^T x & \text{if } Ax \leq b \\ -\infty & \text{if } Ax \not\leq b. \end{cases}$$

$$L_2(y) = \max_{x \geq 0} L(x, y) = \begin{cases} b^T y & \text{if } y^T A = c^T \\ +\infty & \text{if } y^T A \neq c^T. \end{cases}$$

Relative to the canonical form, the optimum value of L_2 is found by solving the linear program

$$\min_{y \geq 0} L_2(y) = \min_{y \geq 0} y^T b \quad \text{s.t.} \quad y^T A = c^T. \tag{33}$$

Nevertheless, it is straightforward to check that Theorem 3.4 is literally valid also for a linear program in its canonical form.

Linear programming problems are particularly important in applications because they can be solved efficiently. In the theory of *cooperative games* with possibly more than two players (see Chapter 8), linear programming is a structurally analytical tool. We do not go into algorithmic details here but refer to the standard mathematical optimization literature.[11]

4.6. *Linear programming games*

A *linear programming game* is a LAGRANGE game that arises from a linear program. If $c \in \mathbb{R}^n, b \in \mathbb{R}^m$ and $A \in \mathbb{R}^{m \times n}$ are the problem parameters, we may denote is by

$$L_+(c, A, b) = \Lambda(\mathbb{R}^n_+, \mathbb{R}^m, L)$$
$$L(c, A, b) = \Lambda(\mathbb{R}^n, \mathbb{R}^m, L),$$

where

$$L(x, y) = c^T x + y^T b - y^T A x$$

is the underlying LAGRANGE function. Linear programming games are convex. In particular, Theorem 3.4 implies:

(1) A linear programming game admits an equilibrium if and only if the underlying linear program has an optimal solution.

(2) The equilibria of a linear programming game are precisely the pairs (x^*, y^*) of primally and dually optimal solutions.

Ex. 3.7. Show that every randomized matrix game is a linear programming game. (See Section 3.3.2.)

---

[11] *e.g.,* FAIGLE *et al.* [17].

Chapter 4

Investing and Betting

The opponent of a gambler is usually a player with no specific optimization goal. The opponent's strategy choices seem to be determined by chance. Therefore, the gambler will have to decide on strategies with good expected returns. Information plays an important role in the quest for the best decision. Hence the problem how to model information exchange and common knowledge among (possibly more than two) players deserves to be addressed as well.

Assume that an *investor* (or *bettor* or *gambler* or simply *player*) is considering a financial engagement in a certain venture. Then the obvious — albeit rather vague — big question for the investor is:

- *What decision should best be taken?*

More specifically, the investor wants to decide whether an engagement is worthwhile at all and, if so, how much of the available capital should be invested how. Obviously, the answer depends on additional information: What is the likelihood of a success? What gain can be expected? What is the risk of a loss? *etc.*

The investor is thus about to participate as a player in a 2-person game with an opponent whose strategies and objective are not always clear or known in advance. Relevant information is not completely (or not reliably) available to the investor so that the decision must be made under uncertainties. Typical examples are gambling and betting where the success of the engagement depends on events that may or may not occur and hence on "fortune" or "chance". But also investments in the stock market fall into this category when it is not clear in advance whether the value of a particular investment will rise or fall.

We are not able to answer the big question above completely but will discuss various aspects of it. Before going into further details, let us illustrate the difficulties of the subject with a classical — and seemingly paradoxical — gambling situation.

The St. Petersburg paradox. Imagine yourself as a potential player in the following game of chance.

Ex. 4.1 (St. Petersburg game). A coin (with faces "H" and "T") is tossed repeatedly until "H" shows. If this happens at the nth toss, a participating player will receive $\alpha_n = 2^n$ euros. There is a participation fee of a_0 euros, however. So the net gain of the player is

$$a = \alpha_n - a_0 = 2^n - a_0$$

if the game stops at the nth toss. At what entrance fee a_0 would a participation in the game be attractive?

Assuming a fair coin in the St. Petersburg game, the probability to go through more than n tosses (and hence to have the first n results as "T") is

$$q_n = \left(\frac{1}{2}\right)^n = \frac{1}{2^n} \to 0 \quad (n \to \infty).$$

So the game ends almost certainly after a finite number of tosses. The expected return to a participant is nevertheless infinite:

$$E_P = \sum_{n=1}^{\infty} 2^n q_n = \frac{2^1}{2^1} + \frac{2^2}{2^2} + \ldots + \frac{2^n}{2^n} + \ldots = +\infty,$$

which might suggest that a player should be willing to pay any finite amount a_0 for being allowed into the game. In practice, however, this could be a risky venture (see Ex. 4.2).

Ex. 4.2. Show that the probability of receiving a return of 100 euros or more in the St. Petersburg game is less than 1%. So a participation fee of $a_0 = 100$ euros or more appears to be not attractive because it will not be recovered with a probability of more than 99%.

Paradoxically, when we evaluate the utility of the return $\alpha_n = 2^n$ not directly but by its logarithm $\log_2 \alpha_n = \log_2 2^n = n$, the St. Petersburg payoff has a finite utility expectation:

$$G_P = \frac{\log_2 2^1}{2^1} + \frac{\log_2 2^2}{2^2} + \ldots + \frac{\log_2 2^n}{2^n} + \ldots = \sum_{n=1}^{\infty} \frac{n}{2^n} < 2.$$

This observation suggests that one should expect an utility value of less than 2 and hence a return of less than $2^2 = 4$ euros.

Remark 4.1 (Logarithmic utilities). The logarithm function as a measure for the utility value of a financial gain was introduced by D. Bernoulli[1] in his analysis of the St. Petersburg game. This concave function plays an important role in our analysis as well.

Whether one uses $\log_2 x$, the logarithm base 2, or the natural logarithm $\ln x$ does not make an essential difference since the two functions differ just by a scaling factor:

$$\ln x = (\ln 2) \cdot \log_2 x.$$

Arithmetic and geometric growth. If investments are repeated, the question arises how to evaluate the evolution of the capital arising from the investments. To make this question more precise, assume that a capital of initial size b_0 takes on the values

$$b_0, b_1, \ldots, b_t, \ldots$$

in discrete time steps. The *arithmetic growth rate up to time t* is the number

$$\alpha_t = \frac{b_t - b_0}{t} \quad \text{and hence} \quad b_t = b_0 + t\alpha_t.$$

---

[1] D. Bernoulli (1700–1782).

The *geometric growth rate up to time t* is the number

$$\gamma_t = \sqrt[t]{b_t/b_0} \quad \text{and hence} \quad b_t = \gamma_t^t \cdot b_0.$$

Passing to logarithms, the geometric growth rate becomes the *logarithmic geometric growth rate*

$$G_t = \ln \gamma_t = \frac{\ln b_n - \ln b_0}{t}, \tag{34}$$

which is the arithmetic growth rate of the logarithmic utility. Since $x \mapsto \ln x$ is a monotonically strictly increasing function, optimizing γ_t is equivalent with optimizing of G_t. In other words:

> Optimizing the geometric growth rate is equivalent with optimizing the arithmetic growth rate of the logarithmic utility.

1. Proportional investing

Our general model consists of a potential investor with an initial portfolio B of $b > 0$ euros (or dollars or...) and an investment opportunity A. If things go well, an investment of size x would bring a return $rx > x$. If things do not go well, the investment will return nothing.

In the analysis, we will denote the *net return rate* by

$$\rho = r - 1.$$

The investor is to decide what portion of B should be invested. The investor believes:

(PI) Things go well with probability $p > 0$ and do not go well with probability $q = 1 - p$.

1.1. *Expected portfolio*

Under the assumption (PI), the investor's expected portfolio value after the investment x is

$$B(x) = [(b - x) + rx]p + (b - x)q = [b + \rho x]p + (b - x)q$$

since an amount of size $b - x$ is not invested and therefore not at risk. The derivative is

$$B'(x) = \rho p - q$$

So $B(x)$ is strictly increasing if $\rho > q/p$ and non-increasing otherwise. Hence, if the investor's decision is motivated by the maximization of the expected portfolio value $B(x)$, the *naive investment rule* applies:

(NIR) If $\rho > q/p$, invest all of B in A and expect the return

$$B(b) = rbp = (1 + q)b > b.$$

If $\rho \leq q/p$, invest nothing since no proper gain is expected.

In spite of its intuitive appeal, rule (NIR) can be quite risky (see Ex. 4.3).

Ex. 4.3. Assume that the investment opportunity A offers the return rate $r = 100$ for an investment in the case it goes well. Assume further that an investor estimates the probability for A to go well to be $p = 10\%$. In view of

$$q/p = 9 \ < \ 99 = r - 1 = \rho,$$

a naive investor would invest the full amount $x^* = b$ and expect a tenfold return:

$$B(x^*) = B(b) = rbp = 10b$$

However, with probability $q = 90\%$, the investment should be expected to result in a total loss.

1.2. Expected utility

With respect to the logarithmic utility function $\ln x$, the expected utility of an investment of size x with net return rate ρ would be

$$U(x) = p \ln(b + \rho x)) + q \ln(b - x).$$

The marginal utility of x is the value of the derivative

$$U'(x) = \frac{\rho p}{b + \rho x} - \frac{q}{b - x}.$$

The second derivative is

$$U''(x) = \frac{\rho^2 p}{(b - \rho x)^2} + \frac{q}{(b - x)^2} > 0.$$

So the investment x^* has optimal (logarithmic) utility if

$$U'(x^*) = 0 \quad \text{or} \quad \frac{\rho p}{b + \rho x^*} = \frac{q}{b - x^*}. \tag{35}$$

Ex. 4.4. In the situation of Ex. 4.3, one has

$$U(x) = \frac{1}{10} \ln(b + 99x) + \frac{9}{10} \ln(b - x)$$

with the derivative

$$U'(x) = \frac{99}{10(b + 99x)} - \frac{9}{10(b - x)}.$$

$U'(x^*) = 0$ implies $x^* = b/11$. Hence the portion $a^* = 1/11$ of the portfolio should be invested in order to maximize the expected utility U. The rest of the portfolio should be retained and not be invested.

The expected value of the portfolio with respect to $x^* = b/11$ is

$$B(x^*) = B(b/11) = (1 + 9/11)b \approx 1.818\, b$$

and thus much lower than under the naive investment policy. However, the investor is *guaranteed* to preserve at least $10/11$ of the original portfolio value.

1.3. *The fortune formula*

Let $a = x/b$ denote the fraction of the portfolio B to be possibly invested into the opportunity A with r-fold return if A works out. With $\rho = r - 1 > 0$, the expected logarithmic utility function then becomes

$$u(a) = U(x/b) = p \ln b(1 + \rho x/b) + q \ln b(1 - x/b)$$

with the derivative

$$u'(a) = \frac{\rho p}{1 + \rho a} - \frac{q}{1 - a}. \tag{36}$$

If a loss is to be expected with positive probability $q > 0$, and the investor decides on a full investment, *i.e.*, chooses $a = 1$, then the expected utility (value)

$$u(1) = \lim_{a \to 0} u(a) = -\infty$$

results — no matter how big the net return rate ρ might be.

On the other hand, the choice $a = 0$ of no investment has the utility

$$u(0) = \ln b.$$

The investment fraction a^* yielding the optimal utility lies somewhere between these extremes.

LEMMA 4.1 *Let $u'(a)$ be as in (36) and $0 < a^* < 1$. Then*

$$u'(a^*) = 0 \quad \Longleftrightarrow \quad a^* = p - q/\rho.$$

Proof. Exercise left to the reader.

The choice of the investment rate a^* with optimal expected logarithmic utility $u(a^*)$ yields the so-called *fortune formula* of KELLY [24]:

$$a^* = p - \frac{q}{\rho} \quad \text{if } 0 < p - \frac{q}{\rho} < 1 \qquad (37)$$

An investment strategy according to (37) is known as a *G-strategy* or KELLY *strategy.*[2]

Betting one's belief. It is important to keep in mind that the probability p in the fortune formula (37) is the subjective evaluation of an investment success by the investor.

The "true" probability is often unknown at the time of the investment. However, if p reflects the investor's best knowledge about the true probability, there is nothing better the investor could do. This truism is known as the investment advice

Bet your belief!

2. Fair odds

An investment into an opportunity A offering a return of $r \geq 1$ euros per euro invested with a certain probability $\Pr(A)$ or returning nothing (with probability $1 - \Pr(A)$) is called a *bet* on A. The investor is then a *bettor* (or a *gambler*) and the net return

$$\rho = r - 1$$

is the payoff of the bet. The payoff is assumed to be guaranteed by a *bookmaker* (or *bank*). The net return rate is also denoted[3] by $\rho : 1$ and known as the *odds* of the bet.

The expected gain (per euro) of the gambler, and hence the bookmaker's loss, is

$$E = \rho \Pr(A) + (-1)(1 - \Pr(A)) = r \Pr(A) - 1.$$

---

[2]See, *e.g.*, ROTANDO AND THORP [38].
[3]The notation differs in various parts of the world: in continental Europe, for example, the gross return rate $r : 1$ is customary.

The odds $\rho : 1$ are considered to be *fair* if the gambler and the bookmaker have the same expected gain, *i.e.*, if

$$E = -E \quad \text{and hence} \quad E = 0$$

holds. In other words:

$$\rho : 1 \quad \text{is fair} \quad \Longleftrightarrow \quad \rho = \frac{1 - \Pr(A)}{\Pr(A)} \quad \Longleftrightarrow \quad r = \frac{1}{\Pr(A)}$$

If the true probability $\Pr(A)$ is not known to the bettor, it needs to be estimated. Suppose the bettor's estimate for $\Pr(A)$ is p. Then the bet appears (subjectively) advantageous if and only if

$$E(p) > 0, \quad \textit{i.e., if } r > 1/p. \tag{38}$$

The bettor will consider the odds $\rho : 1$ as fair if

$$E(p) = 0 \quad \text{and hence} \quad r = 1/p.$$

In the case $E(p) < 0$, of course, the bettor would not expect a gain but a loss on the bet — on the basis of the information that has led to the subjective probability estimate p for $\Pr(A)$.

Examples. Let us look at some examples.

Ex. 4.5 (DE MÉRÉ's game[4]). Let A be the event that *no* "6" shows if a single 6-sided die is rolled four times. Suppose the odds $1 : 1$ are offered on A. If the gambler considers all results as equally likely, the gambler's estimate of the probability for A is

$$p = \frac{5^4}{6^4} = \frac{625}{1296} \approx 0.482 < 0.5$$

because there are $6^4 = 1296$ possible result sequences on 4 rolls of the die, of which $5^4 = 625$ correspond to A. So the player should

---

[4]Mentioned to B. PASCAL (1623–1662).

expect a negative return:

$$E(p) = (\rho + 1)p - 1 = 2p - 1 = 2(5/6)^4 - 1 < 0.$$

In contrast, let $\tilde{A}$ be the event that *no* double 6 shows if a pair of dice is rolled 24 times. Now the prospective gambler estimates $\Pr(\tilde{A})$ as

$$\tilde{p} = (35/36)^{24} > 0.5.$$

Consequently, the odds $1 : 1$ on $\tilde{A}$ would lead the gambler to expect a proper gain:

$$\tilde{E} = 2\tilde{p} - 1 > 0.$$

Ex. 4.6 (Roulette). Let $W = \{0, 1, 2, \ldots, 36\}$ represent a roulette wheel and assume that $0 \in W$ is colored green while eighteen numbers in W are red and the remaining eighteen numbers black. Assume that a number $X \in W$ is randomly determined by spinning the wheel and allowing a ball come to rest at one of these numbers.

(a) Fix $w \in W$ and the gross return rate $r = 18$ on the event

$$A_w = \{X = w\}.$$

Should a gambler expect a positive gain when placing a bet on A_w?

(b) Suppose the bank offers the odds $1 : 1$ on the event $R = \{X = \text{red}\}$. Should a gambler consider these odds on R to be fair?

The doubling strategy. For the game of roulette (see Ex. 4.6) and for similar betting games with odds $1 : 1$ (*i.e.*, with the return rate $r = 2$) a popular wisdom[5] recommends repeated betting according to the following strategy:

---

[5] I have learned strategy (D) myself as a youth from my uncle Max.

(D) Bet the amount 1 on $R = \{X = \text{red}\}$. If R does not occur, continue with the double amount 2 on R. If R does not show, double again and bet 4 on R and so on — until the event R happens.

Once R shows, one has a *net gain* of 1 on the original investment of size 1 (see Ex. 4.7). The probability for R *not* to happen in one spin is $19/37$. So the probability of seeing red in one of the first n spins of an equally balanced roulette wheel is high:

$$1 - (19/37)^n \rightarrow 1 \quad (n \rightarrow \infty).$$

Hence: *Strategy (D) achieves a net gain of 1 with high probability.*

Paradoxically(?), the expected net gain for betting any amount $x > 0$ on the event R is always strictly negative however:

$$E_R = 2x \left(\frac{18}{37}\right) - x = -\frac{x}{37} < 0.$$

Ex. 4.7. Show for the game of roulette with a well-balanced wheel:

(1) If $\{X = \text{red}\}$ shows on the fifth spin of the wheel only, strategy (D) has lost a total of 15 on the first 4 spins. However, having invested 16 more and then winning $2^5 = 32$ on the fifth spin, yields the overall net return

$$32 - (15 + 16) = 1.$$

(2) The probability for $\{X = \text{red}\}$ to happen on the first 5 spins is more than 95%.

REMARK 4.2. The problem with strategy (D) is its risk management. A player has only a limited amount of money available in practice. If the player wants to limit the risk of a loss to B euros, then the number of iterations in the betting sequence is limited to at most k, where

$$2^{k-1} \le B < 2^k \quad \text{and hence} \quad k = \lfloor \log_2 B \rfloor.$$

Consequently:

- The available budget B is lost with probability $(19/37)^k$.
- The portfolio grows to $B + 1$ with probability $1 - (19/37)^k$.

3. Betting on alternatives

Consider k mutually exclusive events $A_0, A_1, \ldots, A_{k-1}$ of which one will occur with certainty and a bank that offers the odds $\rho_i : 1$ for bets on the k events A_i, which means:

> (1) The bank guarantees a total payoff of $r_i = \rho_i + 1$ euros for each euro invested in A_i if the event A_i occurs.
> (2) The bank offers a scenario with $1/r_i$ being the probability for A_i to occur.

Suppose a gambler estimates the events A_i to occur with probabilities p_i and decides to invest the capital B of unit size[6] $b = 1$ fully. Under this condition, a *(betting) strategy* is a k-tuple $a = (a_0, a_1, \ldots, a_{k-1})$ of numbers $a_i \geq 0$ such that

$$a_0 + a_1 + \ldots + a_{k-1} = 1$$

with the interpretation that the portion a_i of the capital will be bet onto the occurrence of event A_i for $i = 0, 1, \ldots, k - 1$. The gambler's expected logarithmic utility of strategy a is

$$U(a, p) = \sum_{i=0}^{k-1} p_i \ln(a_i r_i)$$

$$= \sum_{i=0}^{k-1} p_i \ln a_i + \sum_{i=0}^{k-1} p_i \ln r_i.$$

Notice that $p = (p_0, p_1, \ldots, p_{k-1})$ is a strategy in its own right and that the second sum term in the expression for $U(a, p)$ does not

---

[6]The normalized assumption results in no loss of generality: the analysis is independent of the size of B.

depend on the choice of a. So only the first sum term is of interest when the gambler seeks a strategy with optimal expected utility.

THEOREM 4.1. *Let* $p = (p_0, p_1, \ldots, p_{k-1})$ *be the gambler's probability assessment. Then:*

$$U(a, p) < U(p, p) \iff a \neq p.$$

Consequently, $a^* = p$ *is the strategy with the optimal logarithmic utility under the gambler's expectations.*

Proof. The function $f(x) = x - 1 - \ln x$ is defined for all $x > 0$. Its derivative

$$f'(x) = 1 - 1/x$$

is negative for $x < 1$ and positive for $x > 1$. So $f(x)$ is strictly decreasing for $x < 1$ and strictly increasing for $x > 1$ with the unique minimum $f(1) = 0$. This yields BERNOULLI's *inequality*

$$\ln x \leq x - 1 \quad \text{and} \quad \ln x = x - 1 \iff x = 1. \tag{39}$$

Applying the BERNOULLI inequality, we find

$$U(a, p) - U(p, p) = \sum_{i=0}^{k-1} p_i \ln a_i - \sum_{i=0}^{k-1} p_i \ln p_i = \sum_{i=0}^{k-1} p_i \ln(a_i/p_i)$$

$$\leq \sum_{i=1}^{k-1} p_i(a_i/p_i - 1) = \sum_{i=1}^{k-1} a_i - \sum_{i=1}^{k-1} p_i = 0$$

with equality if and only if $a_i = p_i$ for all $i = 0, 1, \ldots, k_1$. $\qquad \square$

Theorem 4.1 leads to the betting rule with the optimal expected logarithmic utility:

(BR) *For all* $i = 0, 1, \ldots, k - 1$, *bet the portion* $a_i^* = p_i$ *of the capital B on the event A_i.*

REMARK 4.3. The proportional rule (BR) depends *only* on the gambler's probability estimate p. It is independent of the particular odds $\rho_i : 1$ the bank may offer!

Fair odds. As in the proof of Theorem 4.1, one sees:

$$\sum_{i=0}^{k-1} p_i \ln r_i = -\sum_{i=0}^{k-1} p_i \ln(1/r_i) \geq -\sum_{i=0}^{k-1} p_i \ln p_i = \sum_{i=0}^{k-1} p_i \ln(1/p_i)$$

with equality if and only if $r_i = 1/p_i$ holds for all $i = 0, 1, \ldots, k - 1$. It follows that the best odds $\rho_i : 1$ for the bank (and worst for the gambler) are given when

$$\rho_i = r_i - 1 = \frac{1 - p_i}{p_i} \quad (i = 0, 1, \ldots, k - 1). \tag{40}$$

In this case, the gambler expects the logarithmic utility of the optimal strategy p as

$$U(p) = \sum_{i=0}^{k-1} p_i \ln p_i - \sum_{i=0}^{k-1} p_i \ln(1/r_i) = 0.$$

We understand the odds as in (40) to be *fair* in the context of betting with alternatives.

3.1. *Statistical frequencies*

Assume that the gambler of the previous sections has observed:

- The event A_i has popped up s_i times in n consecutive instances of the bet.

So, under the strategy a, the original portfolio B of unit size $b = 1$ would have developed into size

$$B_n(a) = (a_0 r_0)^{s_0} (a_1 r_1)^{s_1} \cdots (a_{k-1} r_{k-1})^{s_{k-1}}$$

with the logarithmic utility

$$U_n(a) = \ln B_n(a) = \sum_{i=0}^{k-1} s_i \ln(a_i r_i)$$

$$= \sum_{i=0}^{k-1} s_i \ln a_i + \sum_{i=0}^{k-1} s_i \ln r_i.$$

As in the proof of Theorem 4.1, the gambler finds in hindsight:

COROLLARY 4.1. *The strategy* $a^* = (s_0/n, \ldots, s_{k-1}/n)$
would have led to the maximal logarithmic utility value

$$U_n(a^*) = \sum_{i=0}^{k-1} s_i \ln(s_i/t) + \sum_{i=0}^{k-1} p_i \ln r_i$$

and hence to the growth with the maximal geometric growth rate

$$B_n(a^*) = \frac{(s_0 r_0)^{s_0}(s_1 r_1)^{s_1} \cdots (s_{k-1} r_{k-1})^{s_{k-1}}}{n^n}.$$

Based on the observed frequencies s_i the gambler might reasonably estimate the events A_i to occur with probabilities according to the relative frequencies

$$p_i = s_i/n \quad (i = 0, 1, \ldots, k-1)$$

and then expect optimal geometric growth.

4. Betting and information

Assuming a betting situation with the k alternatives $A_0, A_1, \ldots,$ A_{k-1} and the odds $\rho_x : 1$ (for $x = 0, 1, \ldots, k-1$) as before, suppose, however, that the event A_x is actually already established — but that the bettor does not have this information before placing the bet.

Suppose further that information now arrives through some (human or technical) *communication channel* K so that the outcome

A_x is reported to the bettor (perhaps incorrectly) as A_y:

$$x \to \boxed{K} \to y.$$

The question is:

- *Having received the ("insider") information "y", how should the bettor place the bet?*

To answer this question, let

$p(x|y)$ = probability for the true result to be x when y is received.

REMARK 4.4. The parameters $p(x|y)$ are typically subjective evaluations of the bettor's trust in the channel K.

A *betting strategy* in this setting of information transmission becomes a $(k \times k)$-matrix A with coefficients $a(x|y) \geq 0$ which satisfy

$$\sum_{x=0}^{k-1} a(x|y) = 1 \quad \text{for} \quad y = 0, 1, \ldots, k-1.$$

According to strategy A, $a(x|y)$ would be the fraction of the budget that is bet on the event A_x when y is received. In particular, the bettor's *trust matrix* P with the coefficients $p(x|y)$ is a strategy. In the case where A_x is the true result, one therefore expects the logarithmic utility

$$U_x(A) = \sum_{y=0}^{k-1} p(x|y) \ln[a(x|y)r_x] = \sum_{y=0}^{k-1} p(x|y) \ln a(x|y) + \ln r_x.$$

As in Corollary 4.1, we find for all $x = 0, 1, \ldots, k-1$:

$$U_x(A) < U_x(P) \quad \Longleftrightarrow \quad a(x|y) = p(x|y) \; \forall y = 0, 1, \ldots, k-1.$$

So the trust matrix P yields also an optimal strategy (under the given trust in K on part of the bettor) and confirms the betting rule:

Bet your belief!

Information transmission. Let $p = (x_0, x_1, \ldots, x_{k-1})$ be the bettor's probability estimates on the k events $A_0, A_1, \ldots, A_{k-1}$ or, equivalently, on the index set $\{0, 1, \ldots, k-1\}$. Then the expected logarithmic utility of strategy A is relative to base 2:

$$U_2^{(p)}(A) = \sum_{x=0}^{k-1}\sum_{y=0}^{k-1} p_x p(x|y) \log_2 a(x|y) + \sum_{x=0}^{k-1} p_x \log_2 r_x.$$

NOTA BENE. *The probabilities p_x are estimates on the likelihood of the events A_x, while the probabilities $p(x|y)$ are estimates on the trust into the reliability of the communication channel K. They are logically not related.*

Setting

$$H(X) = -\sum_{i=1}^{k-1} p_x \log_2 x$$

$$H(r) = -\sum_{i=1}^{k-1} r_x \log_2 x$$

$$H(X, Y) = -\sum_{x=0}^{k-1}\sum_{y=0}^{k-1} p_x p(x|y) \log_2 a(x|y),$$

we thus have

$$U_2^{(p)} = -H(X|Y) - H(r) = U_2(p) + T(X|Y)$$

where

$$T(X|Y) = H(X) - H(X|Y)$$

is the increase of the bettor's expected logarithmic utility due to the communication *via* channel K.

REMARK 4.5 (Channel capacity). Given the channel K as above with transmission (trust) probabilities $p(s|r)$ and the probability distribution

$$p = (p_x | x \in X)$$

on the channel inputs x, the parameter $T(X, Y)$ is the *(information) transmission rate* of K.

Maximizing over all possible input distributions p, one obtains the *channel capacity* $C(K)$ as the smallest upper bound on the achievable transmission rates:

$$C(K) = \sup_p T(X, Y).$$

The parameter $C(K)$ plays an important role in the theory of information and communication in general.[7]

Ex. 4.8. A bettor expects the event A_0 with probability 80% and the alternative event A_1 with probability 20%. What bet should be placed?

Suppose now that an expert tells the bettor that A_1 is certain to happen. What bet should the bettor place under the assumption that the expert is right with probability 90%?

5. Common knowledge

Having discussed information with respect to betting, let us digress a little and take a more general view on information and knowledge.[8] Given a system $\mathfrak{S}$, we ask:

To what extent does common knowledge in a group of agents influence individual conclusions about the state of $\mathfrak{S}$?

To explain more concretely what is meant here, we first discuss a well-known riddle.

---

[7]See SHANNON [42].
[8]More can be found in, *e.g.*, FAGIN *et al.* [10].

5.1. *Red and white hats*

Imagine the following situation:

(I) Three girls, G_1, G_2 and G_3, with *red* hats sit in a circle.
(II) Each girl knows that their hats are either *red* or *white*.
(III) Each girl can see the color of all hats except her own.

Now the teacher comes and announces:

(1) *There is at least one red hat.*
(2) *I will start counting slowly. As soon as someone knows the color of her hat, she should raise her hand.*

What will happen? Does the teacher provide information that goes beyond the common knowledge the girls already have? After all, each girl sees two red hats — and hence *knows* that each of the other girls sees at least one red had as well.

Because of (III), the girls know their hat universe $\mathfrak{H}$ is in one of the 8 states of possible color distributions:

	σ_1	σ_2	σ_3	σ_4	σ_5	σ_6	σ_7	σ_8
G_1	R	R	R	W	R	W	W	W
G_2	R	R	W	R	W	R	W	W
G_3	R	W	R	R	W	W	R	W

None of these states can be jointly ruled out. The entropy H_2^0 of their common knowledge is:

$$H_2^0 = \log_2 8 = 3.$$

The teacher's announcement, however, rules out the state σ_8 and reduces the entropy to

$$H_2^1 = \log_2 7 < H_2^0,$$

which means that the teacher has supplied proper additional information.

At the teacher's first count, no girl can be sure about her own hat because none sees *two* white hats. So no hand is raised, which rules out the states σ_5, σ_6 and σ_7 as possibilities.

Denote now by $P_i(\sigma)$ the set of states thought possible by girl G_i when the hat distribution is actually σ. So we have, for example,

$$P_1(\sigma_3) = \{\sigma_3\}, P_2(\sigma_2) = \{\sigma_2\}, P_3(\sigma_4) = \{\sigma_4\}.$$

Consequently, in each of the states $\sigma_2, \sigma_3, \sigma_4$, at least one girl would raise her hand at the second count and conclude confidently that her hat is *red*, which would signal the state (and hence the hat distribution) to the other girls.

If no hand goes up at the second count, *all* girls know that they are in state σ_1 and will raise their hands at the third count.

In contrast, consider the other extreme scenario and assume:

(I') Three girls, G_1, G_2 and G_3, with *white* hats sit in a circle.

(II) Each girl knows that their hats are either *red* or *white*.

(III) Each girl can see the color of all hats except her own.

The effect of the teacher's announcement is quite different:

- Each girl will immediately conclude that her hat is red and raise her hand because she sees only white hats on the other girls.

This analysis shows:

> (i) The information supplied by the teacher is *subjective*: Even when the information ("there is at least one red hat") is false, the girls will eventually conclude with confidence that they know their hat's color.
>
> (ii) When a girl *thinks* she knows her hat's color, she may nevertheless have arrived at a factually wrong conclusion.

Ex. 4.9. Assume an arbitrary distribution of red and white hats among the three girls. Will the teacher's announcement nevertheless lead the girls to the belief that they know the color of their hats?

5.2. *Information and knowledge functions*

An *event* in the system $\mathfrak{S}$ is a subset $E \subseteq \mathfrak{S}$ of states. We say that the event E *occurs* when $\mathfrak{S}$ is in a state $\sigma \in E$. Denoting by $\mathbf{2}^{\mathfrak{S}}$ the collection of all possible events, we think of a function $P : \mathfrak{S} \to \mathbf{2}^{\mathfrak{S}}$ with the property

$$\sigma \in P(\sigma) \quad \text{for all } \sigma \in \mathfrak{S}$$

as an *information function*. P has the interpretation:

- *If $\mathfrak{S}$ is in the state σ, then P provides the information that the event $P(\sigma)$ has occurred.*

Notice that P is not necessarily a sharp identifier of the "true" state σ: any state $\tau \in P(\sigma)$ is a candidate for the true state under the information function P.

The information function P defines a *knowledge function* $K : \mathbf{2}^{\mathfrak{S}} \to \mathbf{2}^{\mathfrak{S}}$ *via*

$$K(E) = \{\sigma \mid P(\sigma) \subseteq E\}$$

with the interpretation:

- *$K(E)$ is the set of states $\sigma \in \mathfrak{S}$ where P suggests that the event E has certainly occurred.*

LEMMA 4.2. *The knowledge function K of an information function P has the properties:*

(K.1) $K(\mathfrak{S}) = \mathfrak{S}$.
(K.2) $E \subseteq F \implies K(E) \subseteq K(F)$.
(K.3) $K(E \cap F) = K(E) \cap K(F)$.
(K.4) $K(E) \subseteq E$.

Proof. Straightforward exercise, left to the reader. $\square$

Property (K.4) is the so-called *reliability axiom*: If one knows (under K) that E has occurred, then E really has occurred.

Ex. 4.10 (Transparency). Verify the *transparency axiom*

(K.5) $K(K(E)) = K(E)$ for all events E.

Interpretation: *When one knows with certainty that E has occurred, then one knows with certainty that one considers E as having occurred.*

We say that E is *evident* if $E = K(E)$ is true, which means:

• The knowledge function K considers an evident event E as having occurred if and only if E really has occurred.

Ex. 4.11. Show: The set $\mathfrak{S}$ of all possible states constitutes always an evident event.

Ex. 4.12 (Wisdom). Verify the *wisdom axiom*

(K.6) $\mathfrak{S} \setminus K(E) = K(\mathfrak{S} \setminus E)$ for all events E.

Interpretation: *When one does not know with certainty that E has occurred, then one is aware of one's uncertainty.*

5.3. *Common knowledge*

Consider now a set $N = \{p_1, \ldots, p_n\}$ of n players p_i with respective information functions P_i and knowledge functions K_i. We say that the event $E \subseteq \mathfrak{S}$ is *evident* for N if E is evident for each of the members of N, *i.e.*, if

$$E = K_1(E) = \ldots = K_n(E).$$

More generally, an event $E \subseteq \mathfrak{S}$ is said to be *common knowledge* of N in the state σ if there is an event $F \subseteq E$ such that

$$F \text{ is evident for } N \quad \text{and} \quad \sigma \in F.$$

PROPOSITION 4.1. *If the event $E \subseteq \mathfrak{S}$ is common knowledge for the n players p_i with information functions P_i in state σ, then*

$$\sigma \in K_{i_1}(K_{i_2}(\ldots(K_{i_m}(E))\ldots))$$

holds for all sequences $i_1 \ldots i_m$ of indices $1 \leq i_j \leq n$.

Proof. If the event E is common knowledge, it comprises an evident event $F \subseteq E$ with $\sigma \in F$. By definition, we have

$$\in K_{i_1}(K_{i_2}(\ldots(K_{i_m}(F))\ldots)) = F$$

By property (K.2) of a knowledge function (Lemma 4.2), we thus conclude

$$K_{i_1}(K_{i_2}(\ldots(K_{i_m}(E))\ldots)) \supseteq K_{i_1}(K_{i_2}(\ldots(K_{i_m}(F))\ldots)) = F \ni \sigma.$$

$\square$

As an illustration of Proposition 4.1, consider the events

$$K_1(E), K_2(K_1(E)), K_3(K_2(K_1(E))).$$

$K_1(E)$ are all the states where player p_1 is sure that E has occurred. The set $K_2(K_1(E))$ comprises those states where player p_2 is sure that player $p_1(E)$ is sure that E has occurred. In $K_3(K_2(K_1(E)))$ are all the states where player p_3 is certain that player p_2 is sure that player p_1 believes that E has occurred. *And so on.*

5.4. *Different opinions*

Let p_1 and p_2 be two players with information functions P_1 and P_2 relative to a finite system $\mathfrak{S}$ and assume:

- Both players have the same probability estimates $\Pr(E)$ on the occurrence of events $E \subseteq \mathfrak{S}$.

We turn to the question:

- *Can there be common knowledge among the two players in a certain state σ^* that they have* different *likelihood estimates η_1 and η_2 for an event E having occurred?*

Surprisingly(?), the answer can be *"yes"* as Ex. 4.13 shows. For the analysis in the example, recall that the *conditional probability* of an event E given the event A, is

$$\Pr(E|A) = \begin{cases} \Pr(E \cap A)/\Pr(A) & \text{if} \quad \Pr(A) > 0 \\ 0 & \text{if} \quad \Pr(A) = 0. \end{cases}$$

Ex. 4.13. Let $\mathfrak{S} = \{\sigma_1, \sigma_2\}$ and assume $\Pr(\sigma_1) = \Pr(\sigma_2) = 1/2$. Consider the information functions

$$P_1(\sigma_1) = \{\sigma_1\} \quad \text{and} \quad P_1(\sigma_2) = \{\sigma_2\}$$
$$P_2(\sigma_1) = \{\sigma_1, \sigma_2\} = P_2(\sigma_2).$$

For the event $E = \{\sigma_1\}$, one finds

$$\Pr(E|P_1(\sigma_1)) = 1 \quad \text{and} \quad \Pr(E|P_1(\sigma_2)) = 0$$
$$\Pr(E|P_2(\sigma_1)) = 1/2 \quad \text{and} \quad \Pr(E|P_2(\sigma_2)) = 1/2.$$

The ground set $\mathfrak{S} = \{\sigma_1, \sigma_2\}$ corresponds to the event *"the two players differ in their estimates on the likelihood that E has occurred"*. $\mathfrak{S}$ is (trivially) common knowledge in each of the two states σ_1, σ_2.

For a large class of information functions, however, our initial question has the guaranteed answer *"no"*. For an example, let us call an information function P *strict information function* if

(St) Every evident event E is a union of *pairwise disjoint* sets $P(\sigma)$.

PROPOSITION 4.2. *Assume that both information functions P_1 and P_2 are strict. Let $E \subseteq \mathfrak{S}$ be an arbitrary event. Assume that it is common knowledge for the players p_1 and p_1 in state $\sigma^* \in \mathfrak{S}$ that they estimate the likelihood for E having occurred as η_1 resp. η_2. Then necessarily equality $\eta_1 = \eta_2$ holds.*
Hence, if $\eta_1 \neq \eta_2$, the players' likelihood estimates cannot be common knowledge.

Proof. Consider the two events

$$E_1 = \{\sigma \mid \Pr(E|P_1(\sigma)) = \eta_1\}$$
$$E_2 = \{\sigma \mid \Pr(E|P_2(\sigma)) = \eta_2\}.$$

Exactly in the event $E_1 \cap E_2$, player p_1 estimates the probability for the event E having occurred to be η_1 while player p_2's estimate is η_2.

Suppose $E_1 \cap E_2$ is common knowledge in state σ^*, *i.e.*, there exists an event $F \subseteq E_1 \cap E_2$ such that

$$\sigma^* \in F \quad \text{and} \quad K_1(F) = F = K_2(F).$$

Because the information function P_1 is strict, F is the union of pairwise disjoint sets $P_1(\sigma_1), \ldots, P_1(\sigma_k)$, say. Because of $F \subseteq E_1 \cap E_2$, one has

$$\Pr(E|P_1(\sigma_1)) = \cdots = \Pr(E|P_1(\sigma_k)) = \eta_1.$$

Taking Ex. 4.14 into account, we therefore find

$$\Pr(E|F) = \Pr(E|P_1(\sigma_1)) = \eta_1.$$

Similarly, $\Pr(E|F) = \eta_2$ is deduced and hence $\eta_2 = \eta_1$ follows. $\quad\square$

Ex. 4.14. Let A, B be events such that $A \cap B = \emptyset$. Then the conditional probability satisfies:

$$\Pr(E|A) = \Pr(E|B) \quad \Longrightarrow \quad \Pr(E|A \cup B) = \Pr(E|A).$$

Part 3

n-Person Games

Chapter 5

Potentials, Utilities and Equilibria

Before discussing n-person games *per se*, it is useful to go back to the fundamental model of a game Γ being played on a system $\mathfrak{S}$ of states and look at characteristic features of Γ. The aim is a general perspective on the numerical assessment of the value of states and strategic decisions.

1. Potentials and utilities

1.1. *Potentials*

To have a "potential" means to have the capability to enact something. In physics, the term *potential* refers to a characteristic quantity of a system whose change results in a dynamic behavior of the system. Potential energy, for example, may allow a mass to be set into motion. The resulting *kinetic energy* corresponds to the change in the potential. Gravity is thought to result from changes in a corresponding potential, the so-called *gravitational field*, and so on.

Mathematically, a potential is represented as a real-valued numerical parameter. In other words: A *potential* on the system $\mathfrak{S}$ is a function

$$v : \mathfrak{S} \to \mathbb{R}$$

which assigns to a state $\sigma \in \mathfrak{S}$ a numerical value $v(\sigma)$. Of interest is the change in the potential resulting from a state transition $\sigma \to \tau$:

$$\partial v(\sigma, \tau) = v(\tau) - v(\sigma)$$

In fact, up to a constant, the potential $v : \mathfrak{S} \to \mathbb{R}$ is determined by its *marginal potential* $\partial v : \mathfrak{S} \times \mathfrak{S} \to \mathbb{R}$:

LEMMA 5.1 *For any potentials* $v, w : \mathfrak{S} \to \mathbb{R}$, *the two statements are equivalent:*

(1) $\partial v = \partial w$.
(2) *There exists a constant* $K_0 \in \mathbb{R}$ *such that for all* $\sigma \in \mathfrak{S}$,

$$w(\sigma) = v(\sigma) + K_0.$$

Proof. In the case (2), one has

$$\partial(\sigma, \tau) = v(\tau) - v(\sigma) = w(\tau) - w(\sigma) = \partial w(\sigma, \tau)$$

and therefore (1). Conversely, if (1) holds, choose any σ_0 and set $K_0 = w(\sigma_0) - v(\sigma_0)$. Then for all $\sigma \in \mathfrak{S}$, property (2) is apparent:

$$w(\sigma) = w(\sigma_0) + \partial w(\sigma_0, \sigma)$$
$$= K_0 + v(\sigma_0) + \partial v(\sigma_0, \sigma) \; = \; K_0 + v(\sigma). \qquad \square$$

1.2. *Utilities*

A *utility measure* is a function

$$U : \mathfrak{S} \times \mathfrak{S} \to \mathbb{R},$$

which we may interpret as a measuring device for the evaluation of a system transition $\sigma \to \tau$ by a real number $U(\sigma, \tau)$. With U, we associate the so-called *local utility functions* $u^\sigma : \mathfrak{S} \to \mathbb{R}$ with the values

$$u^\sigma(\tau) = U(\sigma, \tau) \quad (\sigma, \tau \in \mathfrak{S}).$$

Hence the utility measure U can be equally well understood as an ensemble of local functions $u^\sigma : \mathfrak{S} \to \mathbb{R}$:

$$U \in \mathbb{R}^{\mathfrak{S} \times \mathfrak{S}} \quad \longleftrightarrow \quad \{u^\sigma \in \mathbb{R}^{\mathfrak{S}} \mid \sigma \in \mathfrak{S}\}$$

Path independence. Given the utility measure U, a path

$$\gamma = (\sigma_0 \to \sigma_1 \to \sigma_2 \to \ldots \to \sigma_{k-1} \to \sigma_k)$$

of system transitions has the aggregated utility weight

$$U(\gamma) = U(\sigma_0, \sigma_1) + U(\sigma_1, \sigma_2) \ldots + U(\sigma_{k-1}, \sigma_k).$$

We say that U is *path independent* if the utility weight of any path depends only on its initial state σ_0 and the final state σ_k:

$$U(\sigma_0 \to \ldots \to \sigma_i \to \ldots \to \sigma_k) = U(\sigma_0, \sigma_k).$$

Ex. 5.1. Show that $U \in \mathbb{R}^{\mathfrak{S} \times \mathfrak{S}}$ is path independent if and only if the aggregated value of every *circuit* (*i.e.*, path that starts and ends in the same state) is zero:

$$U(\sigma_0 \to \sigma_1 \to \ldots \to \sigma_k \to \sigma_0) = 0.$$

In particular, a path-independent U satisfies $U(\sigma, \sigma) = 0$.

PROPOSITION 5.1. *The utility measure $U \in \mathbb{R}^{\mathfrak{S} \times \mathfrak{S}}$ is path independent if and only if $U = \partial u$ holds for some potential $u \in \mathbb{R}^{\mathfrak{S}}$.*

Proof. If U is derived from the potential $u : \mathfrak{S} \to \mathbb{R}$, we have

$$U(\sigma_{i-1}, \sigma_i) = \partial u(\sigma_i, \sigma_{i-1}) = u(\sigma_i) - u(\sigma_{i-1})$$

and, therefore, for any $\gamma = (\sigma_0 \to \sigma_1 \to \ldots \to \sigma_{k-1} \to \sigma_k)$:

$$U(\gamma) = \sum_{i=1}^{k} u(\sigma_i) - \sum_{i=0}^{k-1} u(\sigma_i) = u(\sigma_k) - u(\sigma_0) = U(\sigma_k, \sigma_0),$$

which shows that U is path independent.

Conversely, assume that U is a path independent utility measure. Fix a state σ_0 and consider the potential $u \in \mathbb{R}^{\mathfrak{S}}$ with values

$$u(\sigma) = U(\sigma_0, \sigma).$$

Since U is path independent, we have for all $\sigma, \tau \in \mathfrak{S}$,

$$U(\sigma_0, \sigma) + U(\sigma, \sigma_0) = 0 \quad \text{and hence} \quad U(\sigma, \sigma_0) = -U(\sigma_0, \sigma),$$

which implies

$$U(\sigma, \tau) = U(\sigma, \sigma_0) + U(\sigma_0, \tau) = -u(\sigma) + u(\tau) = \partial u(\sigma, \tau). \quad \square$$

Utility potentials. We refer to $v \in \mathbb{R}^{\mathfrak{S}}$ as the underlying *utility potential* of a context that uses the marginal potential ∂v as its utility measure. Such utility measures are particularly important in application models. The utility measures of cooperative game theory,[1] for example, are typically derived from potentials.

2. Equilibria

When we talk about an "equilibrium" of an utility measure $U \in \mathbb{R}^{\mathfrak{S} \times \mathfrak{S}}$ on the system $\mathfrak{S}$, we make the prior assumption that each state σ has associated a *neighborhood*

$$\mathcal{F}^{\sigma} \subseteq \mathfrak{S} \quad \text{with } \sigma \in \mathcal{F}^{\sigma}$$

and that we concentrate on state transitions to neighbors, *i.e.*, to transitions of type $\sigma \to \tau$ with $\tau \in \mathcal{F}^{\sigma}$.

We now say that a system state $\sigma \in \mathfrak{S}$ is a *gain equilibrium* of U if no feasible transition $\sigma \to \tau$ to a neighbor state τ has a positive utility, *i.e.*, if

$$U(\sigma, \tau) \leq 0 \quad \text{holds for all } \tau \in \mathcal{F}^{\sigma}.$$

Similarly, σ is a *cost equilibrium* if

$$U(\sigma, \tau) \geq 0 \quad \text{holds for all } \tau \in \mathcal{F}^{\sigma}.$$

REMARK 5.1 (Gains and costs). *The negative $C = -U$ of the utility measure U is also a utility measure and one finds:*

σ is a gain equilibrium of U $\Longleftrightarrow$ σ is a cost equilibrium of C

---

[1] *cf.* Chapter 8.

From an abstract point of view, the theory of gain equilibria is equivalent to the theory of cost equilibria.

Many real-world systems appear to evolve in dynamic processes that eventually settle in an equilibrium state (or at least approximate an equilibrium) according to some utility measure. This phenomenon is strikingly observed in physics. But also economic theory has long suspected that economic systems may tend towards equilibrium states.[2]

2.1. *Existence of equilibria*

In practice, the determination of an equilibrium is typically a very difficult computational task. Moreover, many utilities do not even admit equilibria. It is generally not easy just to find out whether an equilibrium for a given utility exists at all. Therefore, one is interested in manageable conditions that allow one to conclude that at least one equilibrium exists.

Utilities from potentials. Consider a utility potential $u : \mathfrak{S} \to \mathbb{R}$ with the marginal utility measure

$$\partial u(\sigma, \tau) = u(\tau) - u(\sigma).$$

Here, the following sufficient conditions offer themselves immediately:

> (1) If $u(\sigma) = \max\limits_{\tau \in \mathfrak{S}} u(\tau)$, then σ is a gain equilibrium.
>
> (2) If $u(\sigma) = \min\limits_{\tau \in \mathfrak{S}} u(\tau)$, then σ is a cost equilibrium.

Since every function on a finite set attains a maximum and a minimum, we find

> **PROPOSITION 5.2.** *If $\mathfrak{S}$ is finite, then every utility potential yields a utility measure with at least one gain and one cost equilibrium.*

---

[2] A.A. COURNOT (1838–1877) [9].

Similarly, we can derive the existence of equilibria in systems that are represented in a coordinate space.

PROPOSITION 5.3. *If $\mathfrak{S}$ can be represented as a compact set $\mathcal{S} \subseteq \mathbb{R}^m$ such that $u : \mathcal{S} \to \mathbb{R}$ is a continuous potential, then u yields a utility measure ∂u with at least one gain and one cost equilibrium.*

Indeed, it is well-known that a continuous function on a compact set attains a maximum and a minimum.

REMARK 5.2. Notice that the conditions given in this section are sufficient to guarantee the existence of equilibria — no matter what neighborhood structure on $\mathfrak{S}$ is assumed.

Convex and concave utilities. If the utility measure U under consideration is not implied by a potential function, not even the finiteness of $\mathfrak{S}$ may be a guarantee for the existence of an equilibrium (see Ex. 5.2).

Ex. 5.2. Give the example of a utility measure U relative to a finite state set $\mathfrak{S}$ with no gain and no cost equilibrium.

We now derive sufficient conditions for utilities U on a system whose set $\mathfrak{S}$ of states is represented by a nonempty convex set $\mathcal{S} \subseteq \mathbb{R}^m$. We say:

- U is *convex* if every local function $u^s : \mathcal{S} \to \mathbb{R}$ is convex.
- U is *concave* if every local function $u^s : \mathcal{S} \to \mathbb{R}$ is concave.

THEOREM 5.1. *Let U be a utility measure with continuous local utility functions $u^s : \mathcal{S} \to \mathbb{R}$ on the nonempty compact set $\mathcal{S} \subseteq \mathbb{R}^m$. Then*

(1) *If U is convex, a cost equilibrium exists.*

(2) *If U is concave, a gain equilibrium exists.*

Proof. Define the function $G : \mathcal{S} \times \mathcal{S} \to \mathbb{R}$ with values

$$G(s, t) = u^s(t) \quad \text{for all } s, t \in \mathcal{S}.$$

Then the hypothesis of the Theorem says that G satisfies the conditions of Corollary A.1 of the Appendix. Therefore, an element $s \in \mathcal{S}$ exists such that

$$u^s(t) = G(s, t) \leq G(s, s) = u^s(s) \quad \text{holds for all } s \in \mathcal{S}.$$

Consequently, s^* is a gain equilibrium of U. (The convex case is proved in the same way.) $\qquad\square$

Chapter 6

n-Person Games

> n-person games generalize 2-person games. Yet, it turns out that the special techniques for the analysis of 2-person games apply in this seemingly wider context as well. Traffic systems, for example, fall into this category naturally.

The model of an *n-person game* Γ assumes the presence of a finite set N with $n = |N|$ elements together with a family

$$\mathcal{X} = \{X_i \mid i \in N\}$$

of n further nonempty sets X_i. The elements $i \in N$ are thought of as *players* (or *agents, etc.*). A member $X_i \in \mathcal{X}$ represents the collection of *resources* (or *actions, strategies, decisions, etc.*) that are available to agent $i \in N$.

A *state* of Γ is a particular selection $\mathbf{x} = (x_i \mid i \in N)$ of individual resources $x_i \in X_i$ by the n agents i. So the collection of all states $\mathbf{x}$ of Γ is represented by the direct product

$$\mathfrak{X} = \prod_{i \in N} X_i.$$

It is further assumed that each player $i \in N$ has an individual *utility function*

$$u_i : \mathfrak{X} \to \mathbb{R}$$

111

by which its individual utility of any $\mathbf{x} \in \mathfrak{X}$ is assessed. The whole context

$$\Gamma = \Gamma(u_i \mid i \in N)$$

now describes the n-person game under consideration.

Ex. 6.1. The matrix game Γ with a row player R and a column player C and the payoff matrix

$$P = \begin{bmatrix} (p_{11}, q_{11}) & (p_{12}, q_{12}) \\ (p_{21}, q_{21}) & (p_{22}, q_{22}) \end{bmatrix} = \begin{bmatrix} (+1, -1) & (-1, +1) \\ (-1, +1) & (+1, -1) \end{bmatrix}$$

is a 2-person game with the player set $N = \{R, C\}$ and the strategy sets $X_R = \{1, 2\}$ and $X_C = \{1, 2\}$. Accordingly, the set of states is

$$\mathfrak{X} = X_R \times X_C = \{(1,1), (1,2), (2,1), (2,2)\}.$$

The individual utility functions $u_R, u_C : \mathfrak{X} \to \mathbb{R}$ take the values

$$u_R(s, t) = p_{st} \quad \text{and} \quad u_C(s, t) = q_{st} \quad \text{for all } (s, t) \in \mathfrak{X}.$$

REMARK 6.1. It is often convenient to label the elements of N by natural numbers and assume $N = \{1, 2, \ldots, n\}$ for simplicity of notation. In this case, a state $\mathbf{x}$ of Γ can be denoted in the form

$$\mathbf{x} = (x_1, x_2, \ldots, x_n) \in X_1 \times X_2 \times \cdots \times X_n.$$

Cooperation. The basic game model with a set N of players is readily generalized to a model where groups of players (and not just individuals) derive a utility value from a certain state $\mathbf{x} \in \mathfrak{X}$. To this end, we call a subset $S \subseteq N$ of players a *coalition* and assume an individual utility function $u_S : \mathfrak{X} \to \mathbb{R}$ to exist for each coalition S.

From an abstract mathematical point of view, however, this generalized model can be treated like a standard $|\mathcal{N}|$-person game, having the set

$$\mathcal{N} = \{S \subseteq N\}$$

of coalitions as its set of "superplayers". In fact, we may allow each coalition S to be endowed with its own set X_S of resources.

In this chapter, we therefore retain the basic model with respect to an underlying set N of players.

Further aspects come to the fore, however, when one asks what the strategic decisions at coalition level mean for the individual players. For example:

- *How should one assess the power of an individual player?*
- *How do coalitions come about?*

A special class of potential n-person games with cooperation, so-called *TU-games*, will be studied in their own right in more detail in Chapter 8.

Probabilistic models. There are many probabilistic aspects of n-person games. One consists in having a probabilistic model for the choice of actions to start with (see Ex. 6.2).

Ex. 6.2 (Fuzzy games). Assume a game Γ where any player $i \in N$ has to decide between two alternatives, say "0" and "1", and chooses "1" with probability x_i. Then Γ is an $|N|$-person game in which each player i has the unit interval

$$X_i = [0,1] = \{x \in \mathbb{R} \mid 0 \le x \le 1\}$$

as its set of resources. A joint strategic choice

$$\mathbf{x} = (x_1, \ldots, x_i, \ldots, x_n) \in [0,1]^N$$

can be interpreted as a "fuzzy" decision to form a coalition $X \subseteq N$:

- Player i will be a member of X with probability x_i.

$\mathbf{x}$ is thus the description of a *fuzzy coalition*. Γ is a *fuzzy cooperative game* in the sense of AUBIN [1].

A further model arises from the *randomization* of an n-person game (see Section 3 below). Other probabilistic aspects of n-person games are studied in Chapter 8 and in Chapter 9.

1. Dynamics of *n*-person games

If the game $\Gamma = (u_i \mid i \in N)$ is played, a game instance yields a sequence of state transitions. The transitions are thought to result from changes in the strategy choices of the players.

Suppose $i \in N$ replaces its current strategy x_i by the strategy $y \in X_i$ while all other players $j \neq i$ retain their choices $x_j \in X_j$. Then a state transition $\mathbf{x} \to \mathbf{y} = \mathbf{x}_{-i}(y)$ results, where the new state has the components

$$y_j = \begin{cases} y & \text{if } j = i \\ x_j & \text{if } j \neq i. \end{cases}$$

Two neighboring states $\mathbf{x}$ and $\mathbf{y}$ differ in at most one component. In particular, $\mathbf{x}_{-i}(x_i) = \mathbf{x}$ holds under this definition and exhibits $\mathbf{x}$ as a neighbor of itself. Let us take the set

$$\mathcal{F}_i(\mathbf{x}) = \{\mathbf{x}_{-i}(y) \mid y \in X_i\}$$

as the *neighborhood* of the state $\mathbf{x} \in \mathfrak{X}$ for the player $i \in N$. So the *neighbors* of $\mathbf{x}$ from i's perspective are those states that could be achieved by i with a change of its current strategy x_i, provided all other players $j \neq i$ retain their current strategies x_j.

The utility functions u_i thus provide the natural utility measure U for Γ with the values

$$U(\mathbf{x}, \mathbf{y}) = u_i(\mathbf{y}) - u_i(\mathbf{x}) \quad \text{for all } i \in N \text{ and } \mathbf{x}, \mathbf{y} \in \mathcal{F}_i(\mathbf{x}). \quad (41)$$

Potential games. The *n*-person game $\Gamma = (u_i \mid i \in N)$ is called a *potential game* if there is a potential $v : \mathfrak{X} \to \mathbb{R}$ such that, for all $i \in N$ and $\mathbf{x}, \mathbf{y} \in \mathcal{F}_i(\mathbf{x})$ the marginal utility change equals the change in the potential:

$$u_i(\mathbf{y}) - u_i(\mathbf{x}) = \partial v(\mathbf{x}, \mathbf{y}) = v(\mathbf{y}) - v(\mathbf{x}). \quad (42)$$

2. Equilibria

An *equilibrium* of $\Gamma = \Gamma(u_i \mid i \in N)$ is an equilibrium of the utility measure U as in (41). The joint strategic choice $\mathbf{x} \in \mathfrak{X}$ is thus a gain equilibrium if no player has an utility incentive to switch to another strategy, *i.e.*,

$$u_i(\mathbf{x}) \geq u_i(\mathbf{y}) \quad \text{holds for } all \ i \in N \text{ and } \mathbf{y} \in \mathcal{F}_i(\mathbf{x}).$$

Completely analogously, a *cost equilibrium* is defined *via* the reverse condition:

$$u_i(\mathbf{x}) \leq u_i(\mathbf{y}) \quad \text{holds for } all \ i \in N \text{ and } \mathbf{y} \in \mathcal{F}_i(\mathbf{x}).$$

Aggregated utilities. There is another important view on equilibria. Given the state $\mathbf{x} \in \mathfrak{X}$, imagine that each player $i \in N$ considers an alternative y_i to its current strategy x_i. The aggregated sum of the resulting utility values is

$$G(\mathbf{x}, \mathbf{y}) = \sum_{i \in N} u(\mathbf{x}_{-i}(y_i)) \quad (\mathbf{y} = (y_i \mid y_i \in X_i)).$$

LEMMA 6.1. $\mathbf{x} \in \mathfrak{X}$ *is a gain equilibrium of* $\Gamma(u_i \mid i \in N)$ *if and only if*

$$G(\mathbf{x}, \mathbf{y}) \leq G(\mathbf{x}, \mathbf{x}) \quad \text{holds for all } \mathbf{y} \in \mathfrak{X}.$$

Proof. If $\mathbf{x}$ is a gain equilibrium and $\mathbf{y} = (y_i \mid i \in N) \in \mathfrak{X}$, we have

$$u_i(\mathbf{x}) \geq u(\mathbf{x}_{-i}(y_i)) \quad \text{for all } y_i \in X_i,$$

which implies $G(\mathbf{x}, \mathbf{x}) \geq G(\mathbf{x}, \mathbf{y})$. Conversely, if $\mathbf{x}$ is not a gain equilibrium, there is an $i \in N$ and a $y \in X_i$ such that

$$0 < u_i(\mathbf{x}_{-i}(y)) - u_i(\mathbf{x}) = G(\mathbf{x}, \mathbf{x}_i(y)) - G(\mathbf{x}, \mathbf{x}).$$

which means that $\mathbf{y} = \mathbf{x}_{-i}(y) \in \mathfrak{X}$ violates the inequality. $\qquad\square$

Lemma 6.1 reduces the quest for an equilibrium to the quest for a $\mathbf{x} \in \mathfrak{X}$ that maximizes the associated component function $g^{\mathbf{x}} : \mathfrak{X} \to \mathbb{R}$ of the aggregated utility measure G with the values

$$g^{\mathbf{x}}(\mathbf{y}) = G(\mathbf{x}, \mathbf{y}).$$

It follows immediately that we can carry over the general sufficient conditions in Chapter 5 for the existence of equilibria to the n-person game $\Gamma = (u_i \mid i \in N)$ with utility aggregation function G:

> (1) If Γ is a potential game with a finite set $\mathfrak{X}$ of states, then the existence of a gain and of a cost equilibrium is guaranteed.
> (2) If $\mathfrak{X}$ is represented as a nonempty compact and convex set in a finite-dimensional real parameter space, and all the maps $\mathbf{y} \mapsto G(\mathbf{x}, \mathbf{y})$ are continuous and concave, then Γ admits a gain equilibrium.
> (3) If $\mathfrak{X}$ is represented as a nonempty compact and convex set in a finite-dimensional real parameter space, and all the maps $\mathbf{y} \mapsto G(\mathbf{x}, \mathbf{y})$ are continuous and convex, then Γ admits a cost equilibrium.

Ex. 6.3. Show that the matrix game in Ex. 6.1 is not a potential game. (Hint: The set of states is finite.)

3. Randomization of matrix games

An n-person game $\Gamma = (u_i \mid i \in N)$ is said to be a (*generalized*) *matrix game* if all individual strategy sets X_i are finite.

For a motivation of the terminology, assume $N = \{1, \ldots, n\}$ and think of the sets X_i as index sets for the coordinates of a multidimensional matrix U. A particular index vector

$$\mathbf{x} = (x_1, \ldots, x_i, \ldots, x_n) \in X_1 \times \cdots \times X_i \times \cdots \times X_n \ (= \mathfrak{X})$$

thus specifies a position in U with the n-dimensional coordinate entry

$$U_{\mathbf{x}} = (u_1(\mathbf{x}), \ldots, u_i(\mathbf{x}), \ldots, u_n(\mathbf{x})) \in \mathbb{R}^n.$$

Let us now change the rules of the matrix game Γ in the following way:

> (R) Each player $i \in N$ chooses a probability distribution $p^{(i)}$ on X_i and selects the element $x \in X_i$ with probability $p_x^{(i)}$.

Under rule (R), the players are really playing the related n-person game $\overline{\Gamma} = (\overline{u}_i \mid i \in N)$ with resource sets P_i and utility functions $\overline{u}_i$, where

> (1) P_i is the set of all probability distributions on X_i.
> (2) $\overline{u}_i(p)$ is the expected value of u_i relative to the joint probability distribution $p = (p^{(i)} \mid i \in N)$ of the players.

The n-person game $\overline{\Gamma}$ is the *randomization* of the matrix game Γ. The expected total utility value is

$$\overline{u}_i(p^{(1)}, \ldots, p^{(n)}) = \sum_{x_1 \in X_1} \cdots \sum_{x_n \in X_n} u_i(x_1, \ldots, x_n) p_{x_1}^{(1)} \cdots p_{x_n}^{(n)}.$$

As Ex. 6.1 shows, a (non-randomized) matrix game Γ does not necessarily have equilibria. Notice, on the other hand, that the coordinate product function

$$(t_1, \ldots, t_n) \in \mathbb{R}^n \mapsto t_1 \cdots t_n \in \mathbb{R}$$

is continuous and linear in each variable. Each utility function $\overline{u}_i$ of the randomized game $\overline{\Gamma}$ is a linear combination of such functions and, therefore, also continuous and linear in each variable.

Since linear functions are both concave and convex and the state set

$$\mathfrak{P} = P_1 \times \cdots \times P_n$$

of $\overline{\Gamma}$ is convex and compact, we conclude:

> **THEOREM 6.1 (NASH [30]).** *The randomization $\overline{\Gamma}$ of an n-person matrix game Γ admits both a gain and a cost equilibrium.*

REMARK 6.2. An equilibrium of a randomized matrix game is also known as a NASH[1] *equilibrium.*

4. Traffic flows

A fundamental model for the analysis of flows in traffic networks goes back to WARDROP [45]. Let us look at a discrete version[2] of it. It is based on a graph $G = (V, E)$ with a (finite) set V of nodes and set E of (directed) edges e between nodes,

$$\text{v} \xrightarrow{e} \text{w},$$

representing directed connections from nodes to other nodes. The model assumes:

> (W) There is a set N of players. A player $i \in N$ wants to travel along a path in G from a starting point s_i to a destination t_i and has a set $\mathcal{P}_i$ of paths to choose from.

Game-theoretically speaking, a strategic action s of player $i \in N$ means a particular choice of a path $P \in \mathcal{P}_i$. Let us identify a path $P \in \mathcal{P}_i$ with its incidence vector in $\mathbb{R}^E$ with the components

$$P_e = \begin{cases} 1 & \text{if } P \text{ passes through } e \\ 0 & \text{otherwise.} \end{cases}$$

The joint travel path choice **s** of the players generates the *traffic flow*

$$\mathbf{x}^{\mathbf{s}} = \sum_{i \in N} \sum_{P \in \mathcal{P}_i} \lambda_P^{\mathbf{s}} P \quad \text{of size} \quad |\mathbf{x}^{\mathbf{s}}| = \sum_P \lambda_P^{\mathbf{s}} \leq n,$$

where $\lambda_P^{\mathbf{s}}$ is the number of players that choose path P in **s**. The component $x_e^{\mathbf{s}}$ of $\mathbf{x}^{\mathbf{s}}$ is the amount of traffic on edge e caused by the choice **s**.

---

[1] J.F. NASH (1928–2015).
[2] Due to ROSENTHAL [36].

We assume that a traffic flow $\mathbf{x}$ produces congestion costs $c(x_e)$ along the edges e and hence results in the total congestion cost

$$C(\mathbf{x}) = \sum_{e \in E} c_e(x_e) x_e$$

across all edges. An individual player i has the congestion cost just along its chosen path P:

$$C(P, \mathbf{x}) = \sum_{e \in P} c_e(x_e).$$

If we associate with the flow $\mathbf{x}$ the potential of aggregated costs

$$\Phi(\mathbf{x}) = \sum_{e \in E} \sum_{t=1}^{x_e} c_e(t),$$

we find that player i's congestion cost along path P in $\mathbf{x}$ equals the marginal potential:

$$C(P, \mathbf{x}) = \sum_{e \in P} c_e(x_e) = \Phi(\mathbf{x}) - \Phi(\mathbf{x} - P) = \partial_P \Phi(\mathbf{x} - P).$$

It follows that the players in the WARDROP traffic model play an n-person potential game on the finite set $\mathfrak{X}$ of possible traffic flows.

$\mathbf{x} \in \mathfrak{X}$ is said to be a NASH *flow* if no player i can improve its congestion cost by switching from the current path $P \in \mathcal{P}_i$ to the use of another path $Q \in \mathcal{P}_i$. In other words, the NASH flows are the cost equilibrium flows. Since the potential function Φ is defined on a finite set, we conclude

> • The WARDROP traffic flow model admits a NASH flow.

BRAESS' paradox. If one assumes that traffic in the WARDROP model eventually settles in a NASH flow, *i.e.*, that the traffic flow evolves toward a congestion cost equilibrium, the well-known observation of BRAESS [6] is counter-intuitive:

> (B) *It can happen that a reduction of the congestion along*
> *a particular connection* increases(!) *the total congestion*
> *cost.*

As an example of BRAESS' paradox, consider the network $G = (V, E)$ with

$$V = \{s, r, q, t\} \quad \text{and} \quad E = \{(s, r), (s, q), (r, t), (q, t), (r, q)\}.$$

Assume that the cost functions on the edges are

$$c_{sr}(x) = x, c_{sq}(x) = 4, c_{rt}(x) = 4, c_{qt}(x) = x, c_{rq} = 10$$

and that there are four network users, which choose individual paths from the starting point s to the destination t and want to minimize their individual travel times.

Because of the high congestion cost, no user will travel along (r, q). As a consequence, a NASH flow will have two users of path $P = (s \to r \to t)$ while the other two users would travel along $\tilde{P} = (s \to q \to t)$.

The overall cost is:

$$C(2P + 2\tilde{P}) = 2 \cdot 2 + 4 \cdot 2 + 4 \cdot 2 + 2 \cdot 2 = 24.$$

If road improvement measures are taken to reduce the congestion on (r, q) to $c'_{rq} = 0$, a user of path P can lower its current cost $C(P) = 6$ to $C(Q) = 5$ by switching to the path

$$Q = (s \to r \to q \to t).$$

The resulting traffic flow, however, causes a higher overall cost:

$$C'(P + Q + 2\tilde{P}) = 2 \cdot 2 + 4 \cdot 1 + 3 \cdot 1 + 4 \cdot 2 + 3 \cdot 2 = 25.$$

Chapter 7

Potentials and Temperature

> The temperature of a system depends on the measuring device in use, which is mathematically represented as a potential function. BOLTZMANN's approach to the notion of temperature in statistical thermodynamics extends to general systems. Of particular interest are n-person matrix games where the temperature reflects the activity of the player set as a whole with respect to the total utility. The interpretation of the activity as a METROPOLIS process moreover indicates how the strategic decisions of individual players influence the expected value of the measuring device.

Consider a finite system $\mathfrak{S}$ that is in a state σ with probability π_σ. Then $\mathfrak{S}$ has the entropy

$$H(\pi) = \sum_{\sigma \in \Sigma} \pi_\sigma \ln(1/\pi_\sigma) = -\sum_{\sigma \in \Sigma} \pi_\sigma \ln \pi_\sigma.$$

The expected value of a potential $v \in \mathbb{R}^{\mathfrak{S}}$ will be

$$E(v, \pi) = \sum_{\sigma \in \mathfrak{S}} v_\sigma \pi_\sigma.$$

Let us think of v as a numerical measuring device for a certain characteristic feature of $\mathfrak{S}$. In a physical model, the number v_σ could describe the level of inherent "energy" of $\mathfrak{S}$ in the state σ, for example. In economics, the function $v : \mathfrak{S} \to \mathbb{R}$ could be a representative statistic for the general state of the economy. In the context of an n-person game, v_σ could possibly measure a degree of "activity" of the set N of players in the state σ, *etc.*

Of course, the activity level v_σ depends on the particular characteristic feature that is investigated under v. Different features of $\mathfrak{S}$ may display different activity levels in the same state σ.

1. Temperature

If the expected value $\mu = E(v, \pi)$ is a measure for the degree of activity of $\mathfrak{S}$ relative to the probability distribution π and the potential v, there may nevertheless be other probability distributions π' with the same expected value

$$E(v, \pi) = \mu = E(v, \pi').$$

The idea is now to derive a canonical probability distribution β with a given expected value μ. To this end, we select β as the distribution with the largest entropy:

$$H(\beta) = \max\{H(\pi) \mid E(v, \pi) = \mu\}.$$

As it turns out (Lemma 7.1 below), every other probability distribution π yielding the same expected value μ of v will have a strictly smaller entropy $H(\pi) < H(\beta)$, which means that its specification would require more information. In this sense, β is the unique "freest" (*i.e.*, least biased) distribution with expectation μ.

1.1. *BOLTZMANN distributions*

Given the potential $v : \mathfrak{S} \to \mathbb{R}$ with values $v_\sigma = v(\sigma)$, any $t \in \mathbb{R}$ defines a related BOLTZMANN[1] *(probability) distribution* $\beta(t)$ on $\mathfrak{S}$ with the components

$$\beta_\sigma(t) = \frac{e^{v_\sigma t}}{Z(t)} \quad \text{where} \quad Z(t) = \sum_{\sigma \in \mathfrak{S}} e^{v_\sigma t} > 0.$$

REMARK 7.1. The distributions $\beta(t)$ are also known as BOLTZMANN-GIBBS[2] distributions. The function $Z(t)$ is the associated so-called *partition function*.

---

[1] L. BOLTZMANN (1844–1906).
[2] J.W. Gibbs (1839–1903).

Computing derivatives, one finds[3] that the expected value of v relative to $b(t)$ can be expressed as the logarithmic derivative of the partition function:

$$\mu(t) = \sum_{\sigma \in \mathfrak{S}} v_\sigma \beta_\sigma(t) = \frac{Z'(t)}{Z(t)} = \frac{d \ln Z(t)}{dt}. \qquad (43)$$

NOTA BENE. *Under a* BOLTZMANN *distribution* $\beta(t)$ *with* $t \in \mathbb{R}$, *no state of* $\mathfrak{S}$ *will be impossible (i.e., occur with probability* 0) *and no state will occur with certainty (i.e., with probability* 1) *if* $\mathfrak{S}$ *has more than one state.*

In the special case $t = 0$, one has $Z(0) = |\mathfrak{S}|$. So $\beta(0)$ is the uniform distribution on $\mathfrak{S}$ with the average potential value as its expectation:

$$\mu(0) = \frac{1}{|\mathfrak{S}|} \sum_{\sigma \in \mathfrak{S}} v_\sigma.$$

Moreover, one observes the limiting behavior

$$\lim_{t \to -\infty} \mu(t) = \min_{\sigma \in \mathfrak{S}} v_\sigma \quad \text{and} \quad \lim_{t \to +\infty} \mu(t) = \max_{\sigma \in \mathfrak{S}} v_\sigma. \qquad (44)$$

In fact, one finds that any value μ between the minimum and the maximum of v can be achieved as the expected value relative to a unique BOLTZMANN distribution. Moreover, the BOLTZMANN distribution is the one with the largest entropy relative to a prescribed expected value:

LEMMA 7.1. *Let* $v_* = \min_{\sigma \in \mathfrak{S}}$ *and* $v^* = \max_{\sigma \in \mathfrak{S}} v_\sigma$. *Then:*

(0) $v_* = v^* \implies \mu(t) = v^*(= v_*)$ *holds for all* $t \in \mathbb{R}$.
(1) *For every* $v_* < \mu < v^*$, *there exists a unique* $t \in \mathbb{R}$ *such that* $\mu = \mu(t)$.
(2) $H(\pi) < H(\beta(t))$ *holds for every probability distribution* $\pi \neq \beta(t)$ *on* $\mathfrak{S}$ *mit expectation* $E(v, \pi) = \mu(t)$.

---

[3]See also Section 6 of the Appendix.

1.2. *BOLTZMANN temperature*

From Lemma 7.1, it is clear that one could characterize the expected value μ of a non-constant potential v equally well by specifying the parameter $t \in \mathbb{R} \cup \{-\infty, +\infty\}$ of the BOLTZMANN distribution $\beta(t)$ with expectation

$$\mu(t) = \mu.$$

In analogy with the BOLTZMANN model in statistical thermodynamics for the temperature, we call the related parameter

$$T = 1/t.$$

the *temperature* of the system $\mathfrak{S}$ relative to a potential with the expected value $\mu(1/T)$. Adjusting the notation accordingly to

$$\beta^{(T)} = \beta(1/T) \quad \text{and} \quad \mu^{(T)} = \mu(1/T).$$

the BOLTZMANN distribution $\beta^{(T)}$ has the coefficients

$$\beta_\sigma^{(T)} = \frac{e^{v_\sigma/T}}{\sum_{\tau \in \mathfrak{S}} e^{v_\tau/T}} \quad (\sigma \in \mathfrak{S}).$$

As the system "freezes" to the temperature $T = 0$, one obtains the extreme values of the potential v as the expectations in the limit, depending on whether the limit 0 is approached from the positive or the negative side:

$$\lim_{T \to 0^+} \mu^{(T)} = \max_{\sigma \in \mathfrak{S}} v_\sigma$$

$$\lim_{T \to 0^-} \mu^{(T)} = \min_{\sigma \in \mathfrak{S}} v_\sigma.$$

In contrast, all states of $\mathfrak{S}$ are equally likely at when the temperature T is infinite.

2. The METROPOLIS process

METROPOLIS *et al.* [29] have pointed out that BOLTZMANN distributions may result dynamically as limiting distributions of quite general stochastic processes.

To formulate the process, we associate with each $\sigma \in \mathfrak{S}$ a neighborhood $\mathcal{F}(\sigma) \subseteq \mathfrak{S}$ that renders $\mathfrak{S}$ *connected* in the following sense:

- For any $\sigma, \tau \in \mathfrak{S}$, there are $\sigma_1, \ldots, \sigma_k \in \mathfrak{S}$ such that

$$\sigma_{\ell+1} \in \mathcal{F}(\sigma_\ell) \quad \text{holds for all } \ell = 0, \ldots k,$$

where $\sigma_0 = \sigma$ and $\sigma_{k+1} = \tau$.

The METROPOLIS *process* specifies a parameter $t \in \mathbb{R}$ and proceeds from an initial state σ_0 in a (possibly infinite) sequence of state transitions relative to the potential v in the following way.

When the process has reached the state σ, a neighbor $\tau \in \mathcal{F}(\sigma)$ is selected at random as a candidate for a transition $\sigma \to \tau$ and then proceeds as follows:

(1) If $v_\tau \geq v_\sigma$, then the transition $\sigma \to \tau$ is made with certainty, *i.e.*, with probability

$$q_{\sigma\tau} = 1.$$

(2) If $v_\tau < v_\sigma$, then the transition $\sigma \to \tau$ is made with probability

$$q_{\sigma\tau} = e^{(v_\tau - v_\sigma)t} < 1$$

and repeated otherwise. Hence the transition selection procedure is repeated with probability $1 - q_{\sigma\tau}$.

It follows that the METROPOLIS state transitions define a MARKOV chain[4] on $\mathfrak{S}$ with transition probabilities

$$p_{\sigma\tau} = \frac{q_{\sigma\tau}}{|\mathcal{F}(\sigma)|}.$$

The METROPOLIS process converges as a MARKOV chain to a limiting distribution on $\mathfrak{S}$ under quite general conditions. A sufficient condition is, for example:

PROPOSITION 7.1. *Assume that all neighborhoods $\mathcal{F}(\sigma)$ have the same size $f = |\mathcal{F}(\sigma)|$. Then the METROPOLIS process converges to the BOLTZMANN probability distribution $\beta(t)$ in the limit.*

In other words: *The process will arrive at the state $\sigma \in \mathfrak{S}$ after n iterations with probability*

$$p_\sigma^{(n)} \to \beta_\sigma(t) \quad (as\ n \to \infty).$$

Proof. By Ex. A.12 of Section 7 of the Appendix, it suffices to check that the condition

$$\beta_\sigma(t)p_{\sigma\tau} = \beta_\tau(t)p_{\tau\sigma}.$$

is satisfied by any $\sigma, \tau \in \mathfrak{S}$. So assume $v_\tau < v_\sigma$, for example. Then

$$\beta_\sigma(t)p_{\sigma\tau} = \frac{e^{v_\sigma t}e^{(v_\tau - u_\sigma)t}}{Z(t)f} = \frac{e^{v_\tau t}}{Z(t)f} = \beta_\tau(t)p_{\tau\sigma}. \qquad \square$$

Simulated annealing. The METROPOLIS process suggests a simple intuitive method for maximizing a function $v : X \to \mathbb{R}$ over a finite set X:

$$\max_{x \in X} v_x.$$

---

[4]See Section 7 in the Appendix.

(S$_0$) Associate with each $x \in X$ a neighborhood set $\mathcal{F}(x)$ and designate an initial element $x_0 \in X$.
(S$_1$) Choose a sequence $t_0, t_1, \ldots, t_n, \ldots$ of natural numbers t_n with $t_n \to \infty$.
(S$_n$) Iterate the following procedure for $n = 0, 1, 2, \ldots$:

  (1) If $x_n \in X$ has been constructed, choose a neighbor $y \in \mathcal{F}(x_n)$ randomly as a candidate for the next element x_{n+1}.
  (2) If $v_y \geq v_{x_n}$, set $x_{n+1} = y$.
  (3) If $v_y < v_{x_n}$, set
  $$x_{n+1} = \begin{cases} y \text{ with probability } e^{(v_y - v_{x_n})t_n} \\ x_n \text{ with probability } 1 - e^{(v_y - v_{x_n})t_n}. \end{cases}$$

There is no guarantee that the simulating annealing heuristic will find a maximizer of v. However, the analogy with the METROPOLIS procedure makes it intuitively plausible that the method might generate elements x_n with the property

$$v_{x_n} \approx \mu(t_n) \quad \to \quad \max_{x \in X} v_n \quad (n \to \infty).$$

The schematic description of the simulated annealing heuristic leaves many implementation details vague. For example:

- What is the best neighborhood structure in (S$_0$)?
- How should the numbers $t_0, t_1, \ldots$ in (S$_1$) be chosen best?

Nevertheless, the method often produces quite good results in practice[5] and has become a standard in the tool box for discrete optimization problems.

REMARK 7.2. The terminology "annealing" (= cooling) derives from the interpretation of the related parameters $T_n = 1/t_n$ as temperatures which simulate a cooling schedule $T_n \to 0$.

---

[5] See, *e.g.*, S. KIRKPATRICK *et al.* [26].

REMARK 7.3. The import of the simulated annealing method is not so much the generation of BOLTZMANN distributions at fixed temperature levels but the convergence of the expected potential values $\mu(t)$ to the extremum:

$$\mu(t) \;\to\; \max_{\sigma \in \mathfrak{S}} v_\sigma \quad (\text{as } t \to \infty).$$

The latter property can be guaranteed under much more general conditions than are formulated in Proposition 7.1.[6]

3. Temperature of matrix games

Let $\Gamma = \Gamma(u_i \mid i \in N)$ be an n-person game with player set $N = \{1, \ldots, n\}$ where each player $i \in N$ has a finite set X_i of strategic resources and a utility function

$$u_i : \mathfrak{X} \to \mathbb{R} \quad (\text{with } \mathfrak{X} = X_1 \times X_2 \times \ldots \times X_n).$$

In the model of randomized matrix games, it is assumed that the players i choose probability distributions $\pi^{(i)}$ on their strategy sets X_i *independently from each other* and then select elements $x_i \in X_i$ according to those distributions.

Let us drop the stochastic independence assumption and consider the more general model where the joint strategy

$$\mathbf{x} = (x_1, x_2, \ldots, x_n) \in \mathfrak{X}$$

would be chosen by the player set N with a certain probability $\pi_{\mathbf{x}}$. The aggregated total utility value is then expected to be

$$\mu = \sum_{\mathbf{x} \in \mathfrak{X}} \sum_{i \in N} u_i(\mathbf{x}) \pi_{\mathbf{x}}.$$

The players' total utility

$$u(\mathbf{x}) = \sum_{i \in N} u_i(\mathbf{x})$$

---

[6]See, *e.g.*, U. FAIGLE and W. KERN [14].

is a potential on $\mathfrak{X}$. So one may consider the (BOLTZMANN) temperature relative to u. In the case

$$\mu = \frac{1}{Z_T} \sum_{\mathbf{x} \in \mathfrak{X}} e^{u(\mathbf{x})/T} \quad (\text{with } Z_T = Z(1/T))$$

we say that Γ is *is played at temperature* T. If $|T| \approx \infty$ (*i.e.*, $|T|$ is very large), we expect about the average value of the total utility:

$$\mu \approx \frac{1}{|\mathfrak{X}|} \sum_{\mathbf{x} \in \mathfrak{X}} u(\mathbf{x}).$$

If $T > 0$ is very small (*i.e.*, $T \approx 0$), then we may expect about the maximal total utility:

$$\mu \approx \max_{\mathbf{x} \in \mathfrak{X}} u(\mathbf{x}).$$

Similarly, if $T \approx 0$ and $T < 0$ holds, about the minimal total utility value is to be expected:

$$\mu \approx \min_{\mathbf{x} \in \mathfrak{X}} u(\mathbf{x}).$$

A METROPOLIS model. In order to model the dynamic behavior of the player set N as a METROPOLIS process, define the set

$$\mathcal{X} = \{(i, y) \mid i \in N, y \in X_i\}.$$

A selection $(i, y) \in \mathcal{X}$, when the game is in the state $\mathbf{x} \in \mathfrak{X}$, corresponds to a possible transition

$$\mathbf{x} \to \mathbf{x}_{-i}(y)$$

with the interpretation that player i considers the alternative strategy $y \in X_i$ as a candidate for an exchange against the current $x_i \in X_i$. Hence, if (i, y) is chosen randomly, the probability for the corresponding transition is

$$\Pr\{\mathbf{x}_{-i}(y)|\mathbf{x}\} = \frac{1}{|\mathcal{X}|} \min\{1, e^{(u(\mathbf{x}_{-i}(y)) - u(\mathbf{x}))/T}\}.$$

So each $\mathbf{x} \in \mathfrak{X}$ has the same number $f = |\mathcal{X}|$ of neighbors. Clearly, this neighborhood structure is connected. Hence Proposition 7.1

shows that the process converges to a BOLTZMANN probability distribution:

$$p_{\mathbf{x}}^{(n)} \rightarrow \frac{e^{u(\mathbf{x})/T}}{\sum\limits_{\mathbf{y} \in \mathfrak{X}} e^{u(\mathbf{y})/T}}.$$

Recall that every state $\mathbf{x}$ occurs with *strictly positive* probability in a BOLTZMANN distribution at temperature $T \neq 0$. So we should not expect convergence towards an equilibrium with respect to u.

This observation does not contradict the guaranteed existence of an equilibrium in the randomized version of a matrix game (Theorem 6.1). The reason can be found in the different roles of the players in the models:

(1) In a randomized matrix game, each player $i \in N$ chooses, individually, a probability distribution $\pi^{(i)}$ for the selection a $x_i \in X_i$ with the goal of optimizing the individual utility u_i.

(2) In the METROPOLIS process, the total utility u of the group N of players is the key objective.

REMARK 7.4 (Social justice). It appears to be in the joint interest of the community N of players to play Γ at a temperature T that is close to 0 but positive if a large total utility value is desired and negative if a minimal value is sought.

The potential function u is equivalent (up to the scaling factor n) to the average utility function $\overline{u}$ of the members of N:

$$u(\mathbf{x}) = \sum_{i=1}^{n} u_i(\mathbf{x}) \quad \longleftrightarrow \quad \overline{u}(\mathbf{x}) = \frac{1}{n} \sum_{i=1}^{n} u_i(\mathbf{x}).$$

A high group average does not necessarily imply a guaranteed high utility value for *each* individual member in N, however.

To formulate it bluntly:

- Even when a high average utility value is used as a criterion for "social justice" in N, there may still be members of N that are not treated "fairly".

The interplay of different interests (individual utility of the players vs. combined utility of the set of all players) is studied in more details within the framework of *cooperative games* in Chapter 8.

Chapter 8

Cooperative Games

Players in a cooperative game strive for a common goal, from which they possibly profit. Of special interest is the class of TU-games with a transferable utility potential, which is best studied within the context of linear algebra. Central is the question how to distribute the achieved goal's profit appropriately. The core of a cooperative game is an important analytical notion. It strengthens the VON NEUMANN–MORGENSTERN solution concept of stable sets and provides a link to the theory of discrete optimization and greedy algorithms. It turns out that the core is the only stable set in so-called supermodular games. Values of cooperative games are more general solution concepts and can be motivated by stochastic models for the formation of coalitions. Natural models for the dynamics of coalition formation are closely related to thermodynamical models in statistical physics and offer an alternative view on the role of equilibria.

While the agents in the n-person games of the previous chapters typically have individual utility objectives and thus possibly opposing strategic goals, the model of a *cooperative game* refers to a finite set N of $n = |N|$ players that may or may not be active towards a common goal. A subset $S \subseteq N$ of potentially active players is traditionally called a *coalition*. Mathematically, there are several ways of looking at the system of coalitions:

From a set-theoretic point of view, one has the system of the 2^n coalitions

$$\mathcal{N} = \{S \mid S \subseteq N\}.$$

On the other hand, one may represent a subset $S \in \mathcal{N}$ by its incidence vector $x^{(S)} \in \mathbb{R}^N$ with the coordinates

$$x_i^{(S)} = \begin{cases} 1 & \text{if } i \in S \\ 0 & \text{if } i \notin S. \end{cases}$$

The incidence vector $x^{(S)}$ suggests the interpretation of an "activity vector":

$$i \in N \text{ is } active \text{ if } x_i^{(S)} = 1.$$

The coalition S would thus be the collection of active players.

A further interpretation imagines every player $i \in N$ to have a binary strategy set $X_i = \{0,1\}$ from which to choose one element. An incidence vector

$$x = (x_1, \ldots, x_n) \in X_1 \times \cdots \times X_n = \{0,1\}^N \subseteq \mathbb{R}^N$$

represents the joint strategy decision of the n players and we have the correspondence

$$\mathcal{N} \quad \longleftrightarrow \quad \{0,1\}^N = 2^N$$

By a *cooperative game* we will just understand a n-person game Γ with player set N and state set

$$\mathfrak{X} = \mathcal{N} \quad \text{or} \quad \mathfrak{X} = 2^N,$$

depending on a set-theoretic or on a vector space point of view. A general cooperative game $\Gamma = (u_i \mid i \in N)$ with individual utility functions $u_i : \mathcal{N} \to \mathbb{R}$ is therefore a matrix game where each player has the choice between two alternative actions.

In the following, we will concentrate on cooperative games whose individual utilities are implied by a general potential on $\mathcal{N}$.

1. Cooperative TU-games

A *transferable utility* relative to a set N of players is a quantity v whose value $v(S)$ depends on the coalition S of active players and hence is a potential

$$v : \mathcal{N} \to \mathbb{R}.$$

The potential v implies the utility measure $\partial v : \mathcal{N} \times \mathcal{N} \to \mathbb{R}$ with the values

$$\partial v(S, T) = v(T) - v(S).$$

We denote resulting potential game by $\Gamma = (N, v)$ and refer to it as a *cooperative TU-game* with *characteristic function* v.

Ex. 8.1. Assume that the players $i \in N$ evaluate their utility relative to a coalition $S \subseteq N$ by a real parameters $u_i(S)$, which means that each player $i \in N$ has an individual utility function

$$u_i : \mathcal{N} \to \mathbb{R}.$$

The aggregated utility $u = \sum_{i \in N} u_i$ is then a transferable utility and defines the TU-game $\Gamma = (N, u)$. For each $S \subseteq N$ and player $i \in N$, one finds:

$$\partial u_i(S, S \cup \{i\}) = u_i(S \cup \{i\}) - u_i(S)$$
$$= u(S \cup \{i\}) - u(S) = \partial u(S, S \cup \{i\}).$$

REMARK 8.1 (Terminology). Often the characteristic function v of a cooperative TU-game (N, v) is already called a *cooperative game*. In discrete mathematics and computer science a function

$$v : \{0, 1\}^n \to \mathbb{R}$$

is also known as a *pseudo-boolean* function. Decision theory refers to pseudo-boolean functions as *set functions*.[1]

The characteristic function v can represent a *cost* utility or a *profit* utility. The real-world interpretation of the mathematical analysis, of course, depends on whether a cost or a gain model is assumed. Usually, the modeling context makes it clear, however.

---

[1]See, *e.g.*, GRABISCH [21].

Equilibria. Since $\mathcal{N}$ is a finite set, a TU-game (N, v) has equilibria. Indeed, any coalition S with maximal value

$$v(S) = \max_{T \in \mathcal{N}} v(T)$$

is a gain equilibrium, while a minimizer of v is a cost equilibrium.

Normalizations. We will always assume that the characteristic function v of a TU-game is *zero-normalized* in the sense

$$v(\emptyset) = 0. \tag{45}$$

The assumption (45) is made for mathematical convenience. In the case $v(\emptyset) \neq 0$, one may study the related zero-normalized game (N, v') with

$$v'(S) = v(S) - v(\emptyset).$$

Essential features of (N, v) are shared by (N, v'). v and v' imply the same utility measure, for example. In this sense, zero-normalization does not result in a loss of generality in the mathematical analysis.

Other normalizations are sometimes useful. Setting, for example,

$$v''(S) = v(S) - \sum_{i \in S} v(\{i\} \quad \text{for all } S \in \mathcal{N},$$

one obtains a related game (N, v'') where the individual players $i \in N$ have the potential value $v''(\{i\}) = 0$.

In the sequel, we will concentrate on (zero-normalized) TU-games and therefore just talk about a *cooperative game* (N, v).

2. Vector spaces of TU-games

Identifying a TU-game (N, v) with its characteristic function v, we think of the function space

$$\mathbb{R}^{\mathcal{N}} = \{v : \mathcal{N} \to \mathbb{R}\} \quad \text{with} \quad \mathcal{N} = \{S \subseteq N\}$$

as the vector space of all (not necessarily zero-normalized) TU-games on the set N. $\mathbb{R}^{\mathcal{N}}$ is isomorphic with coordinate space $\mathbb{R}^{2^n}$ and has dimension

$$\dim \mathbb{R}^{\mathcal{N}} = |\mathcal{N}| = 2^n = \dim \mathbb{R}^{2^n}.$$

The 2^n unit vectors of $\mathbb{R}^{\mathcal{N}}$ correspond to the so-called DIRAC functions $\delta_S \in \mathbb{R}^{\mathcal{N}}$ with the values

$$\delta_S(T) = \begin{cases} 1 & \text{if } T = S \\ 0 & \text{if } T \neq S. \end{cases}$$

The set $\{\delta_S \mid S \in \mathcal{N}\}$ is a basis of $\mathbb{R}^{\mathcal{N}}$. Any $v \in \mathbb{R}^{\mathcal{N}}$ has the representation

$$v = \sum_{S \in \mathcal{N}} v(S)\delta_S. \tag{46}$$

2.1. *Duality*

It is often advantageous to retain $\mathcal{N}$ as the index set explicitly in the game-theoretic analysis and make use of the set-theoretic structure of the coalition ensemble $\mathcal{N}$. One such example is the *duality operator* $v \mapsto v^*$ on $\mathbb{R}^{\mathcal{N}}$, where

$$v^*(S) = v(N) - v(N \setminus S) \quad \text{for all } S \subseteq N. \tag{47}$$

We say that the game (N, v^*) is the *dual* of (N, v). For any possible coalition $S \in \mathcal{N}$, the numerical value

$$v^*(N \setminus S) = v(N) - v(S) = \partial v(S, N)$$

is the "surplus" of the so-called *grand coalition* N *vs.* S in the game (N, v). So duality expresses a balance

$$v(S) + v^*(N \setminus S) = v(N) \quad \text{for all coalitions } S.$$

Ex. 8.2. Show:

(1) $v \mapsto v^*$ is a linear operator on $\mathbb{R}^{\mathcal{N}}$.
(2) The dual $v^{**} = (v^*)^*$ of the dual v^* of v yields exactly the zero-normalization of v.

2.2. MÖBIUS *transform*

For any $v \in \mathbb{R}^{\mathcal{N}}$, let us define its MÖBIUS[2] *transform* as the function $\hat{v} \in \mathbb{R}^{\mathcal{N}}$ with values

$$\hat{v}(S) = \sum_{T \subseteq S} v(T) \quad (S \in \mathcal{N}). \tag{48}$$

$\hat{v}(S)$ aggregates the v-values of all subcoalitions $T \subseteq S$. In this sense, the MÖBIUS *transformation* $v \mapsto \hat{v}$ is a kind of "discrete integral" on the function space $\mathbb{R}^{\mathcal{N}}$.

Ex. 8.3 (Unanimity games). The MÖBIUS transform $\widehat{\delta}_S$ of the DIRAC function δ_S is known as a *unanimity game* and has the values

$$\widehat{\delta}_S(T) = \begin{cases} 1 & \text{if } S \subseteq T \\ 0 & \text{if } S \not\subseteq T. \end{cases}$$

A coalition T has a non-zero value $\widehat{\delta}_S(T) = 1$ exactly when the coalition T includes *all* members of S. Unanimity games appear to be quite simple and yet are basic (Corollary 8.1 below). Many concepts in cooperative game theory are tested against their performance on unanimity games.

Clearly, the MÖBIUS transformation is a linear operator on $\mathbb{R}^{\mathcal{N}}$. The important observation concerns an inverse property: every characteristic function v arises as the transform of a uniquely determined other characteristic function w.

THEOREM 8.1 (MÖBIUS INVERSION). *For each* $v \in \mathbb{R}^{\mathcal{N}}$, *there is a unique* $w \in \mathbb{R}^{\mathcal{N}}$ *such that* $v = \hat{w}$.

---

[2]A.F. MÖBIUS (1790–1868).

Proof. Recall from linear algebra that, in view of the linearity of the MÖBIUS transformation, it suffices to show:

$$\hat{z} = O \quad \Longrightarrow \quad z = O,$$

i.e., the kernel of the map $v \mapsto \hat{v}$ contains just the zero vector $O \in \mathbb{R}^{\mathcal{N}}$.

So assume that $z \in \mathbb{R}^{\mathcal{N}}$ transforms to $\hat{z} = O$. Let $S \in \mathcal{N}$ be a coalition and observe in the case $S = \emptyset$:

$$z(\emptyset) = \hat{z}(\emptyset) = 0.$$

Assume now, by induction, that $z(T) = 0$ holds for all $T \in \mathcal{N}$ of size $|T| < |S|$. Then the conclusion

$$z(S) = \hat{z}(S) - \sum_{T \subset S} z(T) = 0 - 0 = 0$$

follows and completes the inductive step of the proof. So $z(S) = 0$ must be true for all coalitions S. □

Since the MÖBIUS operator is linear, Theorem 8.1 implies that it is, in fact, an automorphism of $\mathbb{R}^{\mathcal{N}}$, which maps bases onto bases. In particular, we find:

COROLLARY 8.1 (Unanimity Basis). *The unanimity games $\widehat{\delta}_S$ form a basis of $\mathbb{R}^{\mathcal{N}}$, i.e., each $v \in \mathbb{R}^{\mathcal{N}}$ admits a unique representation of the form*

$$v = \sum_{S \in \mathcal{N}} \lambda_S \widehat{\delta}_S \quad \text{with coefficients} \quad \lambda_S \in \mathbb{R}.$$

Ex. 8.4 (HARSANYI dividends). Where $v = \hat{w}$, the values $w(S)$ are known as the HARSANYI *dividends* of the coalitions S in the game (N, v). It follows that the value $v(S)$ of any coalition S is the sum of the HARSANYI dividends of its subcoalitions T:

$$v(S) = \hat{w}(S) = \sum_{T \subseteq S} w(T).$$

REMARK 8.2. The literature is not quite clear on the terminology and often refers to the inverse transformation $\hat{v} \mapsto v$ as the MÖBIUS transformation. Either way, the MÖBIUS transformation is a classical and important tool also in in number theory and in combinatorics.[3]

2.3. *Potentials and linear functionals*

A potential $f : \mathcal{N} \to \mathbb{R}$, interpreted as a vector $f \in \mathbb{R}^{\mathcal{N}}$ defines a linear functional $\tilde{f} : \mathbb{R}^{\mathcal{N}} \to \mathbb{R}$ with the values

$$\tilde{f}(g) = \langle f|g \rangle = \sum_{S \in \mathcal{N}} f_S g_S \quad \text{for all } g \in \mathbb{R}^{\mathcal{N}}.$$

If $g^{(S)}$ is the $(0, 1)$-incidence vector of a particular coalition $S \in \mathcal{N}$, we have

$$\tilde{f}(g^{(S)}) = \langle f|g^{(S)} \rangle = f_S \cdot 1 = f_S,$$

which means that $\tilde{f}$ extends the potential f on 2^N $(= \mathcal{N})$ to all of $\mathbb{R}^{\mathcal{N}}$.

Conversely, every linear functional $g \mapsto \langle f|g \rangle$ on $\mathbb{R}^{\mathcal{N}}$ defines a unique potential f on $\mathcal{N}$ *via*

$$f(S) = \langle f|g^{(S)} \rangle \quad \text{for all } S \in \mathcal{N}.$$

These considerations reveal characteristic functions on $\mathcal{N}$ and linear functionals on $\mathbb{R}^{\mathcal{N}}$ to be two sides of the same coin. From the point of view of linear algebra, one can therefore equivalently define:

> • *A cooperative TU-game is a pair $\Gamma = (N, v)$, where N is a set of players and $v \mapsto \langle v|g \rangle$ is a linear functional on the vector space $\mathbb{R}^{\mathcal{N}}$.*

---

[3]See, *e.g.*, ROTA [37].

2.4. *Marginal values*

The characteristic function v of the cooperative game (N, v) implies the utility measure[4]

$$\partial v(S, T) = v(T) - v(S).$$

Individual players $i \in N$ will assess their value with respect to v by evaluating the change in v that they can effect by being active or inactive.

For a player $i \in N$, we thus obtain its *marginal value* with respect to the coalition S as

$$\partial_i v(S) = \begin{cases} v(S \cup i) - v(S) & \text{if } i \in N \setminus S \\ v(S \setminus i) - v(S) & \text{if } i \in S. \end{cases}$$

Ex. 8.5 (Symmetric difference). The *symmetric difference* of two sets S, T is the set

$$S \Delta T = (S \setminus T) \cup (T \setminus S) = (S \cup T) \setminus (S \cap T).$$

With this notion, the marginal value of player $i \in N$ in (N, v) becomes

$$\partial_i(S) = v(S \Delta i) - v(S) = \partial v(S, S \Delta i).$$

3. Examples of TU-games

3.1. *Additive games*

The marginal value $\partial_i v(S)$ of a player i depends on the coalition S it refers to. Different coalitions may yield different marginal values for the player i.

The TU-game (N, v) is said to be *additive* if a player would add the same marginal value to each coalition it joins. This means: There

---

[4]See also Ex. 8.1.

is a number v_i for each $i \in N$ such that for all coalitions $S \subseteq N$ with $i \notin S$,

$$\partial_i v(S) = v(S \cup i) - v(S) = v_i.$$

Hence, if v is zero-normalized and additive, one has

$$v(S) = \sum_{i \in S} v_i.$$

Conversely, every vector $a \in \mathbb{R}^N$ defines a unique zero-normalized additive game (N, a) with the understanding

$$a(\emptyset) = 0 \quad \text{and} \quad a(S) = \sum_{s \in S} a_i \quad \text{for all } S \neq \emptyset. \tag{49}$$

Ex. 8.6. Which unanimity games (see Ex. 8.3) are additive? Show that the vector space of all additive games on N has dimension $|N| + 1$. The subspace of all zero-normalized additive games on N has dimension $|N|$.

3.2. *Production games*

Similar to the situation in Section 4.3, consider a set N of players in an economic production environment where there are m raw materials, $M_1, \ldots, M_m$ from which goods of k different types may be manufactured.

Let $x = (x_1, \ldots, x_k)$ be a plan that proposes the production of $x_j \geq 0$ units of the jth good and assume:

(1) x would need $a_i(x)$ units of material M_i for all $i = 1, \ldots, m$;
(2) each supplier $s \in N$ has $b_{is} \geq 0$ units of material M_i at its disposal;
(3) the production x could be sold for the price of $f(x)$.

So the coalition $S \subseteq N$ could guarantee a production of market value

$$v(S) = \max_{x \in \mathbb{R}^k_+} f(x) \quad \text{s.t.} \quad a_i(x) \leq \sum_{s \in S} b_{is} \ (i = 1, \ldots, m). \tag{50}$$

The corresponding cooperative TU-game (N, v) is a *production game*.

What is the worth of a player? This is one of the central questions in cooperative game theory. In the context of the production game (N, v), one natural approach to resolve this question is the market price principle:

(MP)   Assuming that each material M_i has a market price of y_i per unit, assign to each supplier $s \in N$ the market value w_s of its inventory:

$$w_s = \sum_{i=1}^{m} y_i b_{is}.$$

An objection against a simple application of the principle (MP) could possibly be made if

$$v(S) > \sum_{s \in S} w_s \quad \text{holds for some coalition } S \subseteq N.$$

In this case, S could generate a market value that is strictly larger than the market value of its inventory. So the intrinsic economic value of the members of S is actually larger than the value of their inventory. This consideration leads to another worth assessment principle:

(CA)   Assign numbers w_s to the members of N such that

$$v(N) = \sum_{s \in N} w_s \quad \text{and} \quad v(S) \le \sum_{s \in S} w_s \text{ for all } S \subseteq N.$$

An allocation $w \in \mathbb{R}^N$ according to principle (CA) is a so-called *core allocation*. Core allocations do not necessarily exist in a given cooperative game, however.[5]

As it turns out, the principles (MP) and (CA) can be satisfied simultaneously if the production game (N, v) has a linear objective and linear restrictions.

---

[5] Core allocations are studied more generally in Section 5.

Linear production games. Assume that the production game with characteristic function (50) is *linear* in the sense

$$f(x) = c^T x = c_1 x_1 + \ldots + c_n x_k$$
$$a_i(x) = a_i^T x = a_{i1} x_1 + \ldots + a_{in} x_k \ (i = 1, \ldots, m)$$

and admits an optimal production plan x^* with market value

$$v(N) = f(x^*) = c^T x^*.$$

x^* is the solution of a linear program. So also an optimal solution y^* exists for the dual linear program

$$\min_{y \in \mathbb{R}_+^m} \sum_{i=1}^m b_i^N y_i \quad \text{s.t.} \quad \sum_{i=1}^m a_{ij} y_i \geq c_j \quad (j = 1, \ldots, k),$$

where we have used the notation for the aggregated inventory of the members of a coalition:

$$b_i^S = \sum_{s \in S} b_{is} \quad \text{for any } S \subseteq N.$$

The components y_i^* of y^* are the shadow prices of the materials M_i. According to principle (MP), let us allocate the individual worth

$$w_s^* = \sum_{i=1}^m y_i^* b_{is} \quad \text{to any } s \in N.$$

To see that w^* satisfies also the principle (CA), observe first from linear programming duality:

$$\sum_{s \in N} w_s^* = \sum_{i=1}^m b_i^N y_i^* = \sum_{j=1}^k c_j x_j^* = v(N).$$

The dual of any S-restricted production problem (50) differs only in the coefficients of the objective function. Expressed in terms of the dual linear program, one has

$$v(S) = \min_{y \in \mathbb{R}_+^m} \sum_{i=1}^m b_i^S y_i \quad \text{s.t.} \quad \sum_{i=1}^m a_{ij} y_i \geq c_j \quad (j = 1, \ldots, k).$$

Since y^* is a feasible (although not necessarily optimal) dual solution for any S-restricted problem, one concludes:

$$v(S) \leq \sum_{i=1}^m b_i^S y_i^* = \sum_{s \in S} w_s^*.$$

3.3. Network connection games

Consider a set $N = \{p_1, \ldots, p_n\}$ of users of some public utility[6] that are to be linked, either directly or indirectly (*via* other users), to some supply node p_0. Assume that the cost of establishing a link between p_i with p_j would be c_{ij} (euros, dollars or whatever). The associated cooperative game has the utility function

$$c(S) = \text{minimal cost of connecting just } S \text{ to } p_0.$$

The relevant question is:

- *How much should a user $p_i \in N$ be charged so that a network with the desired connection can be established?*

One possible cost distribution scheme is derived from a construction method for a connection of minimal total cost $c(N)$:

The *greedy algorithm* builds up a chain of coalitions

$$\emptyset = S_0 \subset S_1 \subset S_2 \subset \ldots \subset S_n = N$$

according to the following iterative procedure:

---

[6]Like electricity or water, *etc.*

(G_0) $S_0 = \emptyset$;

(G_j) If S_j has been constructed, choose $p \in N \setminus S_j$ such that $c(S_j \cup p)$ is as small as possible and charge user p the marginal cost

$$\partial_p v(S) = c(S_j \cup p) - c(S_j).$$

(G_n) Set $S_{j+1} = S_j \cup p$ and continue until all users have been charged.

The greedy algorithm makes sure that the user set N in total is charged the minimal possible connection cost:

$$\sum_{j=1}^{n}[c(S_j) - c(S_{j-1})] = c(S_n) - c(S_0) + \sum_{k=1}^{n-1}[c(S_k) - c(S_k)]$$
$$= c(N) - c(\emptyset)$$
$$= c(N).$$

The greedy algorithm is *efficient* in the sense that the total cost $c(N)$ is distributed. Nevertheless, the greedy cost allocation scheme may appear quite "unfair" from the point of view of individual users (see Ex. 8.7).[7]

Ex. 8.7. Consider a user set $N = \{p_1, p_2\}$ with connection cost coefficients $c_{01} = 100$, $c_{02} = 101$ and $c_{12} = 1$. The greedy algorithm constructs the coalition chain

$$\emptyset = S_0 \subset S_1 = \{p_1\} \subset S_2 = \{p_1.p_2\} = N$$

and charges $c(S_1) = 100$ to user p_1, while user p_2's marginal cost is

$$c(S_2) - c(S_1) = 101 - 100 = 1.$$

So p_1 would be to bear more than 99% of the total cost $c(N) = 101$.

---

[7]Game theorists disagree on "the best" network cost allocation scheme.

3.4. *Voting games*

Assume there is a set N of n voters i of not necessarily equal voting power. Denote by w_i the number of votes voter i can cast. Given a threshold w, the associated *voting game*[8] has the characteristic function

$$v(S) = \begin{cases} 1 & \text{if } \sum_{i \in S} w_i \geq w \\ 0 & \text{otherwise.} \end{cases}$$

In the voting context, $v(S) = 1$ has the interpretation that the coalition S has the voting power to make a certain proposed measure pass. Notice that in the case $v(S) = 0$, a voter i with marginal value

$$\partial_i v(S) = v(S \cup i) - v(S) = 1$$

has the power to swing the vote by joining S. The general question is of high political importance:

- *How can (or should) one assess the overall voting power of a voter i in a voting context?*

REMARK 8.3. A popular index for individual voting power is the BANZHAF power index (see Section 8 below). However, there are alternative evaluations that also have their merits. As in the case of network cost allocation, abstract mathematics cannot decide what the "best" method would be.

4. Generalized coalitions and balanced games

Let us assume that the TU-game (N, v) can be played by several coalitions $S \subseteq N$ "simultaneously", requiring an activity level $y_S \geq 0$ from every member $i \in S$ so that no player has to invest more than 100% of its available activity resources in total. With this in mind,

---

[8] Also known as a *threshold game*.

we define a *generalized coalition*[9] to be a nonnegative vector

$$\mathbf{y} = (y_S \mid S \subseteq N) \in \mathbb{R}_+^{\mathcal{N}} \quad \text{s.t.} \quad \sum_{S \ni i} y_S \leq 1 \; \forall i \in N$$

and associate with it the utility value

$$v(\mathbf{y}) = \langle v | \mathbf{y} \rangle = \sum_{S \subseteq N} v(S) y_S.$$

Ex. 8.8. Assume that $\mathbf{y} = (y_S | S \subseteq N)$ is a generalized coalition with binary components $y_S \in \{0, 1\}$. Show that $\mathbf{y}$ is the incidence vector of a family of pairwise disjoint coalitions.

Ex. 8.9 (Fuzzy coalitions). Let $\pi = (\pi_S | S \subseteq N)$ be a probability distribution on the family $\mathcal{N}$ of all coalitions. Then one has $\pi_S \geq 0$ for all $S \in \mathcal{N}$ and

$$\sum_{S \ni i} \pi_S \leq \sum_{S \in \mathcal{N}} \pi_S = 1 \quad \text{for all } i \in N.$$

So π represents a generalized coalition which generalizes the notion of a fuzzy coalition in the sense of Ex. 6.2.

Denote by $\mathcal{Y} \subseteq \mathbb{R}_+^{\mathcal{N}}$ the collection of all generalized coalitions $\mathbf{y}$ and note that $\mathcal{Y}$ is a nonempty, convex and compact set. The optimal utility value $\overline{v}$ is the optimal solution of a feasible linear program:

$$\overline{v} = \max_{\mathbf{y} \in \mathcal{Y}} v(\mathbf{y}) = \max_{\mathbf{y} \in \mathbb{R}_+^{\mathcal{N}}} \sum_{S \subseteq N} v(S) y_S \quad \text{s.t.} \sum_{S \ni i} y_S \leq 1 \; \forall i \in N. \quad (51)$$

Taking $\mathbf{y}^N$ as the vector with components $y_N = 1$ and $y_S = 0$ if $S \neq N$, we see immediately:

$$\overline{v} \geq v(\mathbf{y}^N) = v(N).$$

The game (N, v) is called *(positively) balanced* if equality is achieved:

$$\overline{v} = v(N).$$

---

[9] Also known as a *packing*.

The dual linear program associated with (51) has the same optimal value:

$$\bar{v} = \min_{x \in \mathbb{R}^N_+} \sum_{i \in N} x_i \quad \text{s.t.} \quad \sum_{i \in S} x_i \geq v(S) \ \forall S \subseteq N. \tag{52}$$

Hence linear programming[10] duality yields:

> **THEOREM 8.2 (BONDAREVA [5]).** *For any cooperative game (N, v), the two statements are equivalent:*
>
> (1) (N, v) *is (positively) balanced.*
> (2) *For each $i \in N$ there is a number $x_i \geq 0$ such that*
>
> $$v(N) = \sum_{i \in N} x_i \quad \text{and} \quad \sum_{i \in S} x_i \geq v(S) \quad \text{for all } S \subseteq N.$$

Ex. 8.10. Let (N, v) be a balanced game. Show:

$$v(N) = \max_{S \subseteq N} v(S).$$

Ex. 8.11 (Linear production games). Show that a linear production game is positively balanced if and only if it admits an optimal production plan.

Covers. The generalized coalition $\mathbf{y} = (y_S | S \subseteq N)$ is said to *cover* the set N if equality

$$\sum_{S \ni i} y_S = 1 \quad \text{holds for all elements } i \in N,$$

which means that each agent i's activity resource of unit value 1 is fully used under $\mathbf{y}$. The *covering value* of (N, v) is the number

$$v^c = \max \ \{v(\mathbf{y}) \mid \mathbf{y} \text{ is a cover of } (N, v)\}.$$

---

[10] *cf.* Theorem 3.4.

As in the derivation of Theorem 8.2, we can characterize the covering value by linear programming duality and find

$$v^c = \min_{x \in \mathbb{R}^N} \sum_{i \in N} x_i \quad \text{s.t.} \quad \sum_{i \in S} x_i \geq v(S) \; \forall S \subseteq N. \qquad (53)$$

Ex. 8.12. Prove formula (53).

Clearly, one has $v(N) \leq v^c \leq \overline{v}$. Calling the game (N, v) *strongly balanced* if it yields the equality

$$v(N) = v^c,$$

we therefore find:

PROPOSITION 8.1. *Every positively balanced game is strongly balanced.*

5. The core

5.1. *Stable sets*

Let us generally understand by a *payoff vector* of a cooperative game (N, v) an assignment x of individual values $x_i \in \mathbb{R}$ to the members $i \in N$ and hence a parameter vector $x \in \mathbb{R}^N$. For any coalition $S \subseteq N$, we customarily use the notation

$$x(S) = \sum_{i \in S} x_i$$

for the total payoff to the members of S under x. The payoff x is said to be *feasible*[11] if

$$x(N) \leq v(N).$$

---

[11] Here we assume again that v represents a gain. For a cost potential c, feasibility of an allocation $x \in \mathbb{R}^N$ means $x(N) \geq c(N)$.

Solution concepts. One of the central issues of cooperative game theory is the question of defining (and then finding) appropriate payoffs to the players in a cooperative game (N, v). A concept for such payoffs is a *solution concept*.

The classical solution concept suggested by VON NEUMANN and MORGENSTERN [34] consists in the identification of so-called *stable* sets of payoff vectors.

To make this notion precise, say that $x \in \mathbb{R}^N$ *dominates* $y \in \mathbb{R}^N$ (relative to v) if there is a coalition $S \subseteq N$ such that each of the members of S is served better under x than under y without exceeding the value $v(S)$, *i.e.*,

$$x_i > y_i \quad \text{holds for all } i \in S \text{ and } \quad x(S) \le v(S).$$

In this context, a set $\mathcal{S} \subseteq \mathbb{R}^N$ is called *stable* if

---

(1) All members of $\mathcal{S}$ are feasible.
(2) Every feasible payoff $z \notin \mathcal{S}$ is dominated by at least one $x \in \mathcal{S}$. (Hence $\mathcal{S}$ must be, in particular, nonempty.)
(3) No $x \in \mathcal{S}$ dominates any other $y \in \mathcal{S}$.

EX. 8.13 (VON NEUMANN–MORGENSTERN). Let $N = \{1, 2, 3\}$ and consider the voting game $\Gamma = (N, v)$ with

$$v(S) = \begin{cases} 1 & \text{if } |S| \ge 2 \\ 0 & \text{if } |S| < 2. \end{cases}$$

Show that $\mathcal{S} = \{(1/2, 1/2, 0), (1/2, 0, 1/2), (0, 1/2, 1/2)\}$ is a stable set.

As intuitively attractive as the solution concept of stable sets may appear, there are some practical drawbacks:

- Not every cooperative game admits a stable set.
- Even if stable sets exist, it may be difficult to identify one.

In the sequel, we will discuss a related solution concept. Although it has similar practical drawbacks, its mathematical ramifications are far-reaching and provide an important link to the mathematical theory of discrete optimization.

5.2. *The core*

Say that the payoff $x \in \mathbb{R}^N$ is *coalition rational* in the game (N, v) if each coalition is awarded at least its own value, *i.e.*, if

$$x(S) \geq v(S) \quad \text{holds for all } S \subseteq N.$$

The *core* of a cooperative profit game (N, v) is the set of all feasible coalition rational payoff vectors:

$$\text{core}(v) = \{x \in \mathbb{R}^N \mid x(N) \leq v(N), x(S) \geq v(S)\ \forall S \subseteq N\}.$$

REMARK 8.4 (Efficiency). Note that every payoff vector $x \in \text{core}(v)$ is *efficient* in the sense

$$x(N) = v(N).$$

Ex. 8.14. Let $x, y \in \text{core}(v)$. Then x cannot dominate y because otherwise a coalition S would exist with the property

$$v(S) \geq x(S) > y(S) \geq v(S),$$

which is a mathematical contradiction.

PROPOSITION 8.2. *Let S be an arbitrary stable set of the cooperative game (N, v). Then*

$$\text{core}(v) \subseteq \mathcal{S}.$$

Proof. Suppose to the contrary, that a vector $y \in \text{core}(v) \setminus \mathcal{S}$ exists. Since $\mathcal{S}$ is stable, it contains a payoff $x \in \mathcal{S}$ that dominates y, *i.e*, there exists a coalition $S \subseteq N$ so that

$$v(S) \leq y(S) < x(S) \leq v(S),$$

which is impossible. $\square$

The core of a cost game (N, c) is defined analogously:

$$\text{core}^*(c) = \{x \in \mathbb{R}^N \mid x(N) \geq c(N), x(S) \leq c(S) \,\forall S \subseteq N\}.$$

Every allocation $x \in \text{core}^*(c)$ distributes the cost $c(N)$ among the players $i \in N$ so that no coalition S pays more than its proper cost $c(S)$.

Ex. 8.15. Show for the (zero-normalized) cooperative game (N, v) and its dual (N, v^*):

$$\text{core}(v^*) = \text{core}^*(v).$$

Ex. 8.16. Give an example of a game (N, v) with $\text{core}(v) = \emptyset$.

PROPOSITION 8.3. *Let (N, v) be an arbitrary TU-game. Then:*

(1) $\text{core}(v) \neq \emptyset \quad \Longleftrightarrow \quad (N, v)$ *is strongly balanced.*
(2) *If $v(\{i\}) \geq 0$ holds for all $i \in N$ in the game (N, v), then*

$$\text{core}(v) \neq \emptyset \quad \Longleftrightarrow \quad (N, v) \text{ is positively balanced.}$$

Proof. Exercise left to the reader (*cf.* Theorem 8.2 and Proposition 8.1).

6. Core relaxations

As Ex. 8.16 shows, the core is not a generally applicable concept for "fair" profit (or cost) distributions to the individual players in a cooperative game (N, v) because it may be empty. To deal with this drawback, various ways have been suggested to relax (and to strengthen) the idea of the core.

6.1. *The open core and least cores*

We associate with (N, v) the *open core* as the set

$$\text{core}^o(v) = \{x \in \mathbb{R}^n \mid x(S) \geq v(S) \,\forall S \subseteq N\}.$$

The set $\text{core}^o(v)$ is obviously never empty. Hence, every nonnegative parameter vector $c \in \mathbb{R}_+^N$ yields a feasible linear program

$$\min \; c^T x \quad \text{s.t.} \quad x \in \text{core}^o(v). \tag{54}$$

Notice that the objective function of (54) is bounded from below, because every $x \in \text{core}^o(v)$ satisfies

$$c^T x = \sum_{i \in N} c_i x_i \geq \sum_{i \in N} c_i v(\{i\}).$$

Hence an optimal solution x^* with an optimal value $c^* = c^*(v) = c^T x^*$ exists. We call the set

$$\text{core}(v, c) = \{x \in \text{core}^o(v) \mid c^T x = c^*\} \tag{55}$$

the *least c-core* of (N, v).

Ex. 8.17 (Classical least core). For $c^T = (1, 1, \ldots, 1)$ with components $c_i = 1$, c^* is precisely the covering value v^c of (N, v) (*cf.* Section 4). The set

$$\text{core}(v, v^c) = \{x \in \text{core}^o(v) \mid x(N) = v^c\}$$

is the classical least core of (N, v).[12]

REMARK 8.5. The least core has a natural game-theoretic interpretation: Suppose that a coalition-rational payoff $x \in \mathbb{R}^N$ to the players incurs the cost $c^T x$. Then $\text{core}(v, c)$ is the collection of all cost-minimal coalition-rational payoffs.

6.2. Nuclea

The idea of the least core is a relaxation of the constraint $x(N) = v(N)$ while retaining the other core constraints $x(S) \geq v(S)$.

An alternative approach to a relaxation of the core concept consists in retaining the equality $x(N) = v(S)$ while possibly relaxing the other constraints.

---

[12]Further game-theoretic implications are studied in, *e.g.*, FAIGLE and KERN [15].

To make the idea precise, say that $f \in \mathbb{R}^{\mathcal{N}}$ is a *relaxation vector* if

$$f_\emptyset = 0 = f_N \quad \text{and} \quad f_S \geq 0 \text{ for all coalitions } S \in \mathcal{N}.$$

f is *feasible* for v if there exists some scalar $\epsilon \in \mathbb{R}$ such that

$$C(f, \epsilon) = \text{core}(v - \epsilon f) \neq \emptyset.$$

Hence, if $\text{core}(v) \neq \emptyset$, every relaxation vector f is feasible (with $\epsilon = 0$, for example).

LEMMA 8.1. *Assume that f is a feasible relaxation with $f_S > 0$ for at least one $S \in \mathcal{N}$. Then there exists a scalar $\epsilon_0 \in \mathbb{R}$ such that*

$$C(f, \epsilon) \neq 0 \iff \epsilon \geq \epsilon_0.$$

Proof. Since f is feasible, the linear program

$$\min_{(\epsilon, x)} \epsilon \quad \text{s.t.} \quad x(N) = v(N), \epsilon f_S + x(S) \geq v(S) \; \forall S \in \mathcal{N}$$

is feasible. Moreover, the objective of the linear program is bounded from below since every feasible solution x satisfies

$$v(N) = x(N) = x(S) + x(N \setminus S)$$
$$\geq v(S) + v(N \setminus S) - \epsilon(f_S + f_{N \setminus S})$$

and therefore

$$\epsilon \geq \frac{v(S) + v(N \setminus S) - v(N)}{f_S + f_{N \setminus S}}.$$

So an optimal minimal value ϵ_0 with the desired property exists. $\square$

If it exists, ϵ_0 tightens the relaxation f to the best possible and yields the nonempty core relaxation

$$C_0 = C(f, \epsilon_0) = \text{core}(v - \epsilon_0 f).$$

In the case $C_0 \neq \emptyset$, one may try to tighten the constraints further in the same way. To this end, we define the family

$$\mathcal{B}_0 = \{S \in \mathcal{N} \mid x(S) = x'(S) \ \forall x, x' \in C_0\}.$$

Let $r_0 = r(\mathcal{B}_0)$ be the rank of the matrix of incidence vectors of the members of $\mathcal{B}_0$. Since each incidence vector is n-dimensional and $N \in \mathcal{B}_0$ holds, we have

$$1 \leq r_0 \leq n.$$

LEMMA 8.2. *If $r_0 = n$, then C_0 contains exactly one single vector x^f.*

Proof. $r_0 = n$ means that every $S \in \mathcal{N}$ is a linear combination of the members of $S \in \mathcal{B}_0$. In particular the singleton coalitions $\{i\}$ are such linear combinations. Since the values $x(S)$ are uniquely determined for all $S \in \mathcal{B}_0$, it follows that also the component values $x_i = x(\{i\})$ are uniquely determined for any $x \in C_0$. Hence there can be only one such x. $\qquad\square$

In the case $r_0 = n$, we call the unique vector $x^f \in C_0$ the *f-nucleon* of the game (N, v).

If $r_0 < n$, let $\mathcal{A}_0$ be the family of coalitions S that are linearly independent of $\mathcal{B}_0$. For every $S \in \mathcal{A}_0$,

$$x(S) > v(S) - \epsilon_0 f_S \quad \text{holds for at least one } x \in C_0.$$

Hence there must exist a vector $x \in C_0$ such that

$$x(S) > v(S) - \epsilon_0 f_S \quad \text{holds for all } S \in \mathcal{A}_0.$$

So these inequalities could possibly be tightened even further and one can try to find the best such $\epsilon_1 < \epsilon_0$ as the optimal solution of the linear program

$$\min_{\epsilon, x} \ \epsilon \quad \text{s.t.} \quad x \in C_0, f_S \epsilon + x(S) \geq v(S) \ \forall S \in \mathcal{A}_0. \tag{56}$$

LEMMA 8.3. ϵ_1 *exists if and only if* $f_S > 0$ *holds for some* $S \in \mathcal{A}_0$.

Proof. Completely analogous to the proof of Lemma 8.1 and left to the reader. ☐

If ϵ_1 exists, we set

$$C_1 = C_1(f) = \{x \in \mathbb{R}^N \mid (\epsilon_1, x) \text{ solves (56)}\} \subset C_0(f)$$

and

$$\mathcal{B}_1 = \{S \in \mathcal{N} \mid x(S) = x'(S) \ \forall x \in C_1\}.$$

Since $S \in \mathcal{B}_1$ holds for at least one $S \in \mathcal{A}_0$, we have

$$n \geq r(\mathcal{B}_1) = r_1 \geq r_0 + 1.$$

If $r_1 = n$, $C_1(f)$ contains a singleton vector x^f, which we now term the *f-nucleon* of (N, v).

In the case $r_1 < n$, the set $\mathcal{A}_1$ of coalitions S that are not linearly dependent on $\mathcal{B}_1$ is non-empty. If $f_S > 0$ holds for at least one $S \in \mathcal{A}_1$, we derive the existence of a minimal objective value

$$\epsilon_2 < \epsilon_1 < \epsilon_0$$

for the linear program

$$\min_{\epsilon, x} \ \epsilon \quad \text{s.t.} \quad x \in C_1, f_S \epsilon + x(S) \geq v(S) \ \forall S \in \mathcal{A}_1. \tag{57}$$

in exactly the same way as before. *And so on.*

After $\nu \leq n$ iterations, we arrive at a situation where one of the two possibilities occurs:

(1) $C_\nu(f) = \{x^f\}$. Then the vector x^f is the *f-nucleon* of (N, v).
(2) $f_S = 0$ holds for all $S \in \mathcal{A}_\nu$. Then the set $C_\nu(f)$ is the *f-nucleon* of (N, v).

Ex. 8.18. Show: $\text{core}(v) \neq \emptyset \iff C_\nu(f) \subseteq \text{core}(v)$.

PROPOSITION 8.4. *Let f be a feasible relaxation vector for the game (N, v) and $\mathcal{F} = \{S \in \mathcal{N} \mid f_S > 0\}$. Then the f-nucleon set $C_\nu(f)$ has the property*

(1) $\nu \leq r(\mathcal{F})$.
(2) *If $r(\mathcal{F}) = n$, $C_\nu(f)$ consists of a singleton x^f.*

6.3. *Nucleolus and nucleon*

The *nucleolus* of the game (N, v) introduced by SCHMEIDLER [40] is the f^1-nucleon relative to the relaxation vector f^1 with the unit parameters

$$f_S^1 = 1 \quad \text{for all coalitions } S \neq \emptyset, N.$$

By Proposition 8.4, it is clear that the nucleolus always exists and is a singleton.[13]

The *nucleon*[14] of a game (N, v) with a nonnegative characteristic function is the f^v-nucleon of the game relative to the relaxations

$$x(S) \geq (1 - \epsilon)v(S) \quad \text{for } 0 \leq \epsilon \leq 1$$

i.e., the relaxation with the coefficients $f^v(S) = v(S)$ for $S \neq N$.

The choice $\epsilon = 1$ shows that the nucleon relaxation is feasible. The nucleon is a singleton vector x^v if the (incidence vectors of the) coalitions $S \in \mathcal{N}$ with value $v(S) > 0$ yield a system of full rank n.

6.4. *Excess minimization*

Let f be a relaxation vector for (N, v) and assume[15]:

$$f_S > 0 \quad \text{for all } S \neq \emptyset, N.$$

---

[13] Related solution concepts are studied in MASCHLER *et al.* [28].
[14] See FAIGLE *et al.* [18].
[15] To keep the discussion simple — the general case can be analyzed in the same spirit.

This assumption implies that f is a feasible relaxation and that (N, v) enjoys a unique f-nucleon x^f.

For any $x \in \mathbb{R}^N$ with $x(N) = v(N)$, define the *excess* of a coalition S with $f_S > 0$ as

$$e(x, S) = \frac{x(S) - v(S)}{f_S} .$$

Since f is feasible, at least one such x exists so that $e(x, S) \geq 0$ holds for all excesses. Set

$$\mathcal{F} = \{x \in \mathbb{R}^N \mid x(N) = v(N), e(x, S) \geq 0 \; \forall S \neq \emptyset, N\}.$$

The parameter ϵ_0 yields the minimal possible overall bound for the excess of a vector $x \in \mathcal{F}$ and one has

$$C_0(f) = \{x \in \mathcal{F} \mid e(x, S) \geq \epsilon_0 \; \forall S \neq \emptyset, N\}.$$

One can now continue and determine the members of $C_0(f)$ with minimal excess and the family $\mathcal{A}_0$ of coalitions whose excess is not yet fixed. This yields the next parameter ϵ_1 and the set

$$C_1(f) = \{x \in C_0(f) \mid e(x, S) \geq \epsilon_1 \; \forall S \in \mathcal{A}_0\}.$$

And so on. This shows:

- The f-nucleon x^f of (N, v) arises from a sequence of excess minimizations.

In other words:

> The nucleon x^f is the unique vector that distributes the value $v(N)$ among the members of N so that each coalition ends with the minimal possible excess.

There are further value assignment concepts in cooperative games that are of interest. Section 8 will provide examples of such concepts. For the moment, let us continue with the study of the core in its own right.

7. MONGE vectors and supermodularity

In principle, it suffices to compute the nucleon x^f of a cooperative game relative to a suitable relaxation vector f in order to have a core vector if the core is nonempty at all (*cf.* Ex. 8.18). The nucleon is typically difficult to compute, however. MONGE[16] vectors are based on an alternative approach, which is computationally more direct. Games all of whose MONGE vectors lie in the core turn out to be particularly interesting.

Throughout the discussion in this section, we consider a fixed zero-normalized cooperative game (N, v) with n players and collection $\mathcal{N}$ of coalitions.

7.1. *The* MONGE *algorithm*

Given a parameter vector $c \in \mathbb{R}^N$ and an arrangement $\pi = i_1 i_2 \ldots i_n$ of the elements of N, the MONGE *algorithm* constructs a *primal* MONGE *vector* $x^\pi \in \mathbb{R}^N$ and a *dual* MONGE *vector* $y^\pi \in \mathbb{R}^{\mathcal{N}}$ as follows:

(M₁) Set $S_0^\pi = \emptyset$ and $S_k^\pi = \{i_1, \ldots, i_k\}$ for $k = 1, 2, \ldots, n$.

(M₂) Set $x_{i_k}^\pi = v(S_k^\pi) - v(S_{k-1}^\pi)$ for $k = 1, 2, \ldots, n$.

(M₃) Set $y_{S_n}^\pi = c_{i_n}$ and $y_{S_\ell}^\pi = c_{i_\ell} - c_{i_{\ell+1}}$ for $\ell = 1, 2, \ldots, n-1$.
  Set $y_S^\pi = 0$ otherwise.

Notice that the construction of the primal MONGE vector x^π proceeds like the greedy algorithm for the cost distribution in network connection games of Section 3 relative to the chain

$$\emptyset = S_0^\pi \subset S_1^\pi \subset \ldots \subset S_n^\pi = N.$$

It is not hard to see[17] that the MONGE vectors x^π and y^π satisfy the identity

$$m^\pi(c) = \sum_{i \in N} c_i x_i^\pi = \sum_{S \in \mathcal{N}} v(S) y_S^\pi. \tag{58}$$

---

[16] G. MONGE (1746–1818).

[17] *cf.* Section 5 of the Appendix.

Different arrangements π and ψ of N, of course, may yield different Monge sums $m^\pi(c)$ and $m^\psi(c)$. Important is the following observation.

LEMMA 8.4. *Let* $\pi = i_1 i_2 \ldots i_n$ *and* $\psi = j_1 j_2 \ldots j_n$ *be two arrangements of* N *with non-increasing c-values:*

$$c_{i_1} \geq c_{i_2} \geq \ldots \geq c_{i_n} \quad \text{and} \quad c_{j_1} \geq c_{j_2} \geq \ldots \geq c_{j_n}.$$

Then

$$m^\pi(c) = \sum_{S \in \mathcal{N}} v(S) y_S^\pi = \sum_{S \in \mathcal{N}} v(S) y_S^\psi = m^\psi(c).$$

Proof. Note that $c_{i_\ell} = c_{i_{\ell+1}}$, for example, implies $y_{S_\ell}^\pi = 0$. So we may assume that the components of c have pairwise different values. But then $\pi = \psi$ holds, which renders the claim trivial. □

7.2. *The* MONGE *extension*

Lemma 8.4 says that there is a well-defined real-valued function $c \mapsto [v](c)$ with the values

$$[v](c) = m^\pi(c) \quad \text{for any } \pi = i_1 \ldots i_n \text{ s.t. } c_{i_1} \geq \cdots \geq c_{i_n}. \tag{59}$$

The function $[v] : \mathbb{R}^N \to \mathbb{R}$ is called the MONGE *extension* of the characteristic function $v : \mathcal{N} \to \mathbb{R}$.

To justify the terminology "extension", consider a coalition $S \subseteq N$ and its $(0,1)$-incidence vector $c^{(S)}$ with the component $c_i^{(S)} = 1$ in the case $i \in S$.

An appropriate MONGE arrangement of the elements of N first lists all 1-components and then all 0-components:

$$\pi = i_1 \ldots i_k \ldots i_n \quad \text{s.t.} \quad c_{i_1}^{(S)} \ldots c_{i_k}^{(S)} \ldots c_{i_n}^{(S)} = 11 \ldots 1100 \ldots 00.$$

Hence we have $y_S^\pi = 1$ and $y_T^\pi = 0$ for $T \neq S$ and conclude

$$[v](c^{(S)}) = v(S) \quad \text{for all } S \subseteq N,$$

which means that $\hat{v}$ and v agree on (the incidence vectors of) $\mathcal{N}$.

REMARK 8.6 (CHOQUET [8] and LOVÁSZ [27]).
Applying the idea and construction of the MONGE sums (58) to non-decreasing value arrangements $f_1 \leq \ldots \leq f_n$ of nonnegative functions $f : N \to \mathbb{R}_+$, one arrives at the CHOQUET *integral*

$$\int f dv = \sum_{k=1}^{n} f_k(v(A_k) - v(A_{k+1})),$$

where $A_k = \{k, k+1, \ldots, n\}$ and $A_{n+1} = \emptyset$.

The map $f \mapsto \int f dv$ is the so-called LOVÁSZ *extension* of the function $v : \mathcal{N} \to \mathbb{R}$. Of course, *mutatis mutandis*, all structural properties carry over from MONGE to CHOQUET and LOVÁSZ.

7.3. *Linear programming aspects*

Generalizing the approach to the notion of balancedness of Theorem 8.2, let us consider the linear program

$$\min_{x \in \mathbb{R}^N} c^T x \text{ s.t. } x(N) = v(N), \ x(S) \geq v(S) \text{ if } S \neq N. \tag{60}$$

and its dual

$$\max_{y \in \mathbb{R}^{\mathcal{N}}} v^T y \text{ s.t. } \sum_{S \ni i} y_S \leq c_i \ \forall i \in N, \ y_S \geq 0 \text{ if } S \neq N. \tag{61}$$

for a given parameter vector $c \in \mathbb{R}^N$. Observe in the case

$$c_{i_1} \geq \ldots \ldots \geq c_{i_n}$$

that the dual MONGE vector y^π relative to c is a dually feasible solution since $y_S^\pi \geq 0$ holds for all $S \neq N$. The feasible primal solutions, on the other hand, are exactly the members of core(v).

Hence, if core$(v) \neq \emptyset$, both linear programs have optimal solutions. Linear programming duality then further shows

$$\tilde{v}(c) = \min_{x \in \text{core}(v)} c^T x \geq v^T y^\pi = [v](c). \tag{62}$$

THEOREM 8.3. $\tilde{v} = [v]$ *holds for the game* (N, v) *if and only if all primal* MONGE x^π *vectors lie in* core(v).

Proof. Assume $c_{i_1} \geq \ldots \geq c_{i_n}$ and $\pi = i_1 \ldots i_n$. If $x^\pi \in \text{core}(v)$, then x^π is a feasible solution for the linear program (60). Since the dual MONGE vector y^π is feasible for (61), we find

$$c^T x^\pi \geq \tilde{v}(c) \geq [v](v) = c^T x^\pi \quad \text{and hence} \quad \tilde{v}(c) = [v](c).$$

Conversely, $\tilde{v} = [v]$ means that the dual MONGE vector is guaranteed to yield an optimal solution for (61). So consider an arrangement $\psi = j_1 \ldots j_n$ of N and the parameter vector $c \in \mathbb{R}^N$ with the components

$$c_{j_k} = n + 1 - k \quad \text{for } k = 1, \ldots, n.$$

The dual vector y^ψ has strictly positive components $y^\psi_{S_k} = 1 > 0$ on the sets S^ψ_k. It follows from the KKT-conditions for optimal solutions that an optimal solution $x^* \in \text{core}(v)$ of the corresponding linear program (60) must satisfy the equalities

$$x^*(S^\psi_k) = \sum_{i \in S_k} x^*_i = v(S^\psi_k) \quad \text{for } k = 1, \ldots, n,$$

which means that x^* is exactly the primal MONGE vector x^π and, hence, that $x^\pi \in \text{core}(v)$ holds. □

7.4. *Concavity*

Let us call the characteristic function $v : 2^N \to \mathbb{R}$ *concave* if v arises from the restriction of a concave function to the $(0, 1)$- incidence vector $c^{(S)}$ of the coalitions S, *i.e.*, if there is a concave function $f : \mathbb{R}^N \to \mathbb{R}$ such that

$$v(S) = f(c^{(S)}) \quad \text{holds for all } S \subseteq N.$$

Accordingly, the cooperative game (N, v) is *concave* if v is concave. We will not pursue an investigation of general concave cooperative games here but focus on a particularly important class of concave games which are closely tied to the MONGE algorithm *via* Theorem 8.3.

PROPOSITION 8.5. *If all MONGE vectors of the game (N, v) lie in* $\text{core}(v)$, *then* (N, v) *is concave.*

Proof. By Theorem 8.3, the hypothesis of the Proposition says

$$\tilde{v}(c) = [v](c) \quad \text{for all } c \in \mathbb{R}^N.$$

Consequently, it suffices to demonstrate that $\tilde{v}$ is a concave function. Clearly, $\widetilde{\lambda c} = \lambda \tilde{v}\langle c\rangle$ holds for all scalars $\lambda > 0$, *i.e.*, $\tilde{v}$ is positively homogeneous.

Consider now arbitrary parameter vectors $c, d \in \mathbb{R}^N$ and $x \in$ core(v) such that $\tilde{v}(c + d) = (c + d)^T x$. Then

$$\tilde{v}(c + d) = c^T x + d^T x \geq \tilde{v}(c) + \tilde{v}(d),$$

which exhibits $\tilde{v}$ as concave.   □

REMARK 8.7. The converse of Proposition 8.5 is not true: there are concave games whose core does not include all primal MONGE vectors.

A word of terminological caution. The game-theoretic literature often applies the terminology "convex cooperative game" to games (N, v) having all primal MONGE vectors in core(v). In our terminology, however, such games are not *convex* but *concave*.

To avoid terminological confusion, one may prefer to refer to such games as *supermodular games* (*cf.* Theorem 8.4 below).

7.5. *Supermodularity*

The central notion that connects the MONGE algorithm with concavity is the notion of *supermodularity*:

THEOREM 8.4. *For the cooperative game (N, v), the following statements are equivalent:*

(I) *All MONGE vectors $x^\pi \in \mathbb{R}^N$ lie in* core(v).
(II) *v is* supermodular, *i.e., v satisfies the inequality*

$$v(S \cap T) + v(S \cup T) \geq v(S) + v(T) \quad \text{for all } S, T \subseteq N.$$

Proof. Assuming (I), arrange the elements of N to $i_1 i_2 \ldots i_n$ such that

$$S \cap T = \{i_1, \ldots, i_k\}, S = \{i_1, \ldots, i_\ell\}, S \cup T = \{i_1, \ldots, i_m\}.$$

By the definition of the Monge algorithm, x^π then satisfies

$$x^\pi(S \cap T) = v(S \cap T), x^\pi(S) = v(S), x^\pi(S \cup T) = v(S \cup T).$$

Moreover, $x^\pi(T) \geq v(T)$ holds if $x^\pi \in \text{core}(v)$. So we deduce the supermodular inequality

$$v(S \cap T) + v(S \cup T) = x^\pi(S \cap T) + x^\pi(S \cup T)$$
$$= x^\pi(S) + x^\pi(T)$$
$$\geq v(S) + v(T).$$

Conversely, let $\pi = i_1, \ldots, i_n$ be an arrangement of N. We want to show:

$$x^\pi \in \text{core}(v).$$

Arguing by induction on $n = |N|$, we note $x^\pi(\emptyset) = v(\emptyset)$ and assume:

$$x^\pi(S) \geq v(S) \quad \text{holds for all subsets } S \text{ of } N' = \{i_1, \ldots, i_{n-1}\}.$$

Consequently, if $i_n \in S$, the supermodularity of v yields

$$x^\pi(S) = x^\pi(S \cap N') + x^\pi_{i_n}$$
$$= x^\pi(S \cap N') + v(N) - v(N')$$
$$\geq v(S \cap N') + v(S \cup N') - v(N')$$
$$\geq v(S). \qquad \square$$

Ex. 8.19. The preceding proof uses the fact that any vector $x \in \mathbb{R}^N$ satisfies the *modular equality*

$$x(S \cap T) + x(S \cup T) = x(S) + x(T) \quad \text{for all } S, T \subseteq N.$$

As Shapley[18] has observed, the concepts of core and stable sets coincide in supermodular games.

---

[18]L.S. Shapley (1923–2016).

THEOREM 8.5 (SHAPLEY [44]). *The core of a supermodular game (N, v) is a stable set.*

Proof. We already know that no core vector dominates any other core vector. Let $y \notin \text{core}(v)$ be an arbitrary feasible payoff vector. We claim that y is dominated by some core vector a. Consider the function

$$g(S) = \frac{v(S) - y(S)}{|S|} \quad \text{on the family } \mathcal{N} \setminus \{\emptyset\}.$$

Since $y \notin \text{core}(v)$, the maximum value g^* of g is strictly positive and attained at some coalition Z. Choose an order $\pi = i_1 \ldots i_j \ldots i_n$ of the elements of N such that $Z = \{i_1, \ldots, i_k\}$ and let x^π be the corresponding primal MONGE vector.

Define the payoff vector $a \in \mathbb{R}^N$ with components

$$a_{i_j} = \begin{cases} y_i + g^* & \text{if } j \leq k \\ x_{i_j}^\pi & \text{if } j > k. \end{cases}$$

Because $x^\pi(Z) = v(Z) = kg^* + y(S)$, it is clear that a is feasible and dominates y relative to the coalition Z. It remains to show that a is a core vector.

Keeping in mind that $x^\pi \in \text{core}(v)$ holds (since v is supermodular), we find for any coalition S:

$$\begin{aligned} a(S) &= a(S \cap Z) + a(S \cap (N \setminus Z)) \\ &= y(S \cap Z) + |S \cap Z|g^* + x^\pi(S \cap (N \setminus Z)) \\ &\geq v(S \cap Z) + v(S \cap (N \setminus Z)) \\ &\geq v(S). \end{aligned} \qquad \square$$

REMARK 8.8. Since the core of a supermodular game is stable and hence contains any other stable set, we know that $\text{core}(v)$ is actually the *unique* stable set of the supermodular game (N, v).

7.6. Submodularity

A characteristic function v is called *submodular* if the inequality

$$v(S \cap T) + v(S \cup T) \le v(S) + v(T) \quad \text{holds for all } S, T \subseteq N.$$

Ex. 8.20. Show for the zero-normalized game (N, v) the equivalence of the statements:

(1) v is supermodular.
(2) v^* is submodular.
(3) $w = -v$ is submodular.

In view of the equality $\text{core}(c^*) = \text{core}^*(c)$ (Ex. 8.15), we find that the Monge algorithm also constructs vectors in the $\text{core}^*(c)$ of cooperative cost games (N, c) with submodular characteristic functions c.

Remark 8.9. Note the fine point of Theorem 8.4, which in the language of submodularity says: (N, c) *is a submodular cost game if and only if* all Monge *vectors* x^π *lie in* $\text{core}^*(c)$.

Network connection games are typically *not* submodular. Yet, the particular greedy cost distribution vector discussed in Section 3.3 does lie in $\text{core}^*(c)$, as the ambitious reader is invited to demonstrate.

Remark 8.10. Because of the Monge algorithm, sub- and super-modular functions play a prominent role in the field of discrete optimiziation.[19] In fact, many results of discrete optimization have a direct interpretation in the theory of cooperative games. Conversely, the model of cooperative games often provides conceptual insight into the structure of discrete optimization problems.

Remark 8.11 (Greedy algorithm). The *Monge* algorithm, applied to linear programs with core-type constraints, is also known as the *greedy algorithm* in discrete optimization.

---

[19]See, *e.g.*, S. Fujishige [20].

8. Values

While the marginal value $\partial_i v(S)$ of player i's decision to join respectively to leave the coalition S is intuitively clear, it is less clear how the overall strength of i in a game should be assessed. From a mathematical point of view, there are infinitely many possibilities to do this.

In general, we understand by a *value* for the class of all TU-games (N, v) a vector-valued function

$$\Phi : \mathbb{R}^{\mathcal{N}} \to \mathbb{R}^N$$

that associates with every characteristic function v a vector $\Phi(v) \in \mathbb{R}^N$. Given Φ, the coordinate value $\Phi_i(v)$ is the assessment of the strength of $i \in N$ in the game (N, v) according to the evaluation concept Φ.

8.1. *Linear values*

The value Φ is said to be *linear* if Φ is a linear operator, *i.e.*, if one has for all games v, w and scalars $\lambda \in \mathbb{R}$, the equality

$$\Phi(\lambda v + w) = \lambda \Phi(v) + \Phi(w).$$

In other words: Φ is linear if each component function Φ_i is a linear functional on the vector space $\mathbb{R}^{\mathcal{N}}$.

Recall from Ex. 8.3 that the unanimity games form a basis of $\mathbb{R}^{\mathcal{N}}$, which means that every game v can be uniquely expressed as a linear combination of unanimity games. Hence a linear value Φ is completely determined by the values assigned to unanimity games. The same is true for any other basis of $\mathbb{R}^{\mathcal{N}}$, of course.

Indeed, if $v_1, \ldots, v_k \in \mathbb{R}^{\mathcal{N}}$ are (the characteristic functions of) arbitrary TU-games and $\lambda_1, \ldots, \lambda_k$ arbitrary real scalars, the linearity of Φ yields

$$\Phi(\lambda_1 v_1 + \ldots + \lambda_k v_k) = \lambda_1 \Phi(v_1) + \ldots + \lambda_k \Phi(v_k).$$

We give two typical examples of linear values.

The value of SHAPLEY [43]. Consider the unanimity game $\widehat{\delta}_T$ relative to the coalition $T \in \mathcal{N}$ where

$$\widehat{\delta}_T(S) = \begin{cases} 1 & \text{if } S \supseteq T \\ 0 & \text{otherwise.} \end{cases}$$

In this case, it might appear reasonable to assess the strength of a player $s \in N \setminus T$ as null, *i.e.*, with the value $\Phi_j^{Sh}(\widehat{\delta}_T) = 0$, and the strength of each of the players $t \in T$ in equal proportion as

$$\Phi_t^{Sh}(\widehat{\delta}_T) = \frac{1}{|T|}.$$

Extending Φ^{Sh} to all games v by linearity in the sense

$$\boxed{v = \sum_{T \in \mathcal{N}} \lambda_t \widehat{\delta}_T \implies \Phi^{Sh}(v) = \sum_{T \in \mathcal{N}} \lambda_T \Phi^{Sh}(\widehat{\delta}_T)}$$

one obtains a linear value $v \mapsto \Phi^{Sh}(v)$, the so-called SHAPLEY *value*.

Ex. 8.21. Show that the players in (N, v) and in its zero-normalization (N, v^{**}) are assigned the same SHAPLEY values.

The power index of BANZHAF [2]. The BANZHAF power index Φ^B appears at first sight to be quite similar to the SHAPLEY value, assessing the power value

$$\Phi_s^B(\widehat{\delta}_T) = 0 \quad \text{if } s \in N \setminus T$$

while treating all $t \in T$ as equals. Assuming $T \neq \emptyset$, the mathematical difference lies in the scaling factor:

$$\Phi_i^B(\widehat{\delta}_T) = \frac{1}{2^{|T|-1}} \quad \text{for all } t \in T.$$

As done with the SHAPLEY value, the BANZHAF power index is extended by linearity from unanimity games to all games (N, v) and thus gives rise to a linear value $v \mapsto \Phi^B(v)$.

We will see in Section 8.2 that the difference between the values Φ^{Sh} and Φ^B can also be explained by two different probabilistic assumptions about the way coalitions are formed.

REMARK 8.12 (Efficiency). If $|T| \geq 1$, the SHAPLEY value distributes the total amount

$$\sum_{i \in N} \Phi_i^{Sh}(\widehat{\delta}_T) = 1 = \widehat{\delta}_T(N)$$

to the members of N and is, therefore, said to be *efficient*. In contrast, we have

$$\sum_{i \in N} \Phi_i^{B}(\widehat{\delta}_T) = \frac{|T|}{2^{|T|-1}} < 1 = \widehat{\delta}_T(N) \quad \text{if } |T| \geq 2.$$

So the BANZHAF power index in not efficient.

REMARK 8.13. One can show that linear values for TU-games typically arise as so-called *least square values*, namely as values obtained from least square approximations with additive games under linear constraints.[20]

8.2. *Random values*

The concept of a random value is based on the assumption that a player $i \in N$ joins a coalition $S \subseteq N \setminus \{i\}$ with a certain probability π_S as a new member. The expected marginal value of $i \in N$ thus is

$$E_i^\pi(v) = \sum_{S \subseteq N \in \{i\}} \partial_i v(S) \pi_S.$$

The function $E^\pi : \mathbb{R}^{\mathcal{N}} \to \mathbb{R}^N$ with components $E_i^\pi(v)$ is the associated *random value*.

Notice that marginal values are linear. Indeed, if $u = \lambda v + w$, one has

$$\partial_i u(S) = \lambda \partial_i v(S) + \partial_i w(S)$$

for all $i \in N$ and $S \subseteq N$. Therefore, the random value E^π is linear as well:

$$E^\pi(\lambda v + w) = \lambda E^\pi(v) + E^\pi(w). \tag{63}$$

---

[20] *cf.* FAIGLE and GRABISCH [13].

REMARK 8.14. The linearity relation (63) implicitly assumes that the probabilities π_S are independent of the particular characteristic function v. If π_S depends on v, the linearity of E^π is no longer guaranteed.

The BOLTZMANN value (to be discussed in Section 9 below) is a random value that is *not* linear in the sense of (63) because the associated probability distribution depends on the characteristic function.

8.2.1. *The value of* BANZHAF

As an example, let us assume that a player i joins any of the 2^{n-1} coalitions $S \subseteq N \setminus \{i\}$ with equal likelihood, *i.e.*, with probability

$$\pi_S^B = \frac{1}{2^{n-1}}.$$

Consider the unanimity game $v_T = \widehat{\delta}_T$ and observe that $\partial_i v_T(S) = 0$ holds if $i \notin T$. On the other hand, if $i \in T$, then one has

$$\partial_i v_T(S) = 1 \iff T \setminus \{i\} \subseteq S.$$

So the number of coalitions S with $\partial_i v_T(S) = 1$ equals

$$|\{S \subseteq N \setminus \{i\} \mid T \subseteq S \cup \{i\}\}| = 2^{n-|T|-1}.$$

Hence we conclude

$$E_i^{\pi^B}(v_T) = \sum_{S \subseteq N \setminus \{i\}} \partial_i v_T(S)\pi_S^B = \frac{2^{n-|T|-1}}{2^{n-1}} = \frac{1}{2^{|T|}}, \qquad (64)$$

which means that the random value E^{π^B} is identical with the BANZHAF power index. The probabilistic approach yields the explicit formula

$$\Phi_i^B(v) = E_i^{\pi^B}(v) = \frac{1}{2^{n-1}} \sum_{S \subseteq N \setminus \{i\}} (v(S \cup i) - v(S)) \quad (i \in N). \quad (65)$$

8.2.2. *Marginal vectors and the* SHAPLEY *value*

Let us imagine that the members of N build up the "grand coalition" N in a certain order

$$\sigma = i_1 i_2 \ldots i_n$$

and hence join in the sequence of coalitions

$$\emptyset = S_0^\sigma \subset S_1^\sigma \cdots \subset S_k^\sigma \subset \cdots \subset S_n^\sigma = N$$

where $S_k^\sigma = S_{k-1}^\sigma \cup \{i_k\}$ for $k = 1, \ldots, n$. Given the game (N, v), σ gives rise to the *marginal vector*[21] $\partial^\sigma(v) \in \mathbb{R}^N$ with components

$$\partial_{i_k}^\sigma(v) = v(S_k^\sigma) - v(S_{k-1}^\sigma) \quad (k = 1, \ldots, n).$$

Notice that $v \mapsto \partial^\sigma(v)$ is a linear value by itself. We can randomize this value by picking the order σ from the set Σ_N of all orders of N according to a probability distribution π. Then the expected marginal vector

$$\partial^\pi(v) = \sum_{\sigma \in \Sigma_N} \partial^\sigma(v) \pi_\sigma$$

represents, of course, also a linear value on $\mathbb{R}^\mathcal{N}$.

Ex. 8.22. Show that the value $v \mapsto \partial^\pi(v)$ is linear and efficient. (Hint: Recall the discussion of the greedy algorithm for network connection games.)

THEOREM 8.6. *The* SHAPLEY *value results as the expected marginal vector relative to the uniform probability distribution on* Σ_N, *where all orders are equally likely:*

$$\Phi^{Sh}(v) = \frac{1}{n!} \sum_{\sigma \in \Sigma_N} \partial^\sigma(v).$$

(Recall from combinatorics that there are $n! = |\Sigma_N|$ *ordered arrangements of* N.)

---

[21] The marginal vectors are precisely the primal MONGE vectors.

Proof. Because of linearity, it suffices to prove the Proposition for unanimity games $v_T = \hat{\delta}_T$. For any order $\sigma = i_1 \ldots i_n \in \Sigma_N$ and element $i_k \in N \setminus T$, we have $\partial_{i_k}(v_T) = 0$ and hence

$$\frac{1}{n!} \sum_{\sigma \in \Sigma_N} \partial_i^\sigma(v_T) = 0 \quad \text{for all } i \in N \setminus T.$$

On the other hand, the uniform distribution treats all members $i \in T$ equally and thus distributes the value $v_T(N) = 1$ equally and efficiently among the members of T:

$$\frac{1}{n!} \sum_{\sigma \in \Sigma_N} \partial_i^\sigma(v_T) = \frac{v_T(N)}{|T|} = \frac{1}{|T|} \quad \text{for all } i \in T,$$

which is exactly the concept of the SHAPLEY value. $\square$

COROLLARY 8.2. *The* SHAPLEY *value* Φ^{Sh} *satisfies:*

(1) $\Phi^{Sh}(v) \in \mathrm{core}(v)$ *if v is supermodular.*
(2) $\Phi^{Sh}(v) \in \mathrm{core}^*(v)$ *if v is submodular.*

Proof. A marginal vector is a primal MONGE vector relative to the zero vector $c = O \in \mathbb{R}^N$. So all marginal vector lie in $\mathrm{core}(v)$ if v is supermodular. Since $\mathrm{core}(v)$ is a convex set and $\Phi^{Sh}(v)$ a convex linear combination of marginal vectors, $\Phi^{Sh}(v) \in \mathrm{core}(v)$ follows. $\square$

We may interpret the SHAPLEY value within the framework of random values initially introduced. To this end, we assume that an order $\sigma \in \Sigma_N$ is chosen with probability $1/n!$ and consider a coalition $S \subseteq N \setminus \{i\}$. We ask:

- What is the probability π_S^{Sh} that i would join S?

Letting $k - 1 = |S|$ be the size of S, the number of sequences σ where i would be added to S is

$$|\{\sigma \mid i = i_k \text{ and } S_{k-1}^\sigma = S\}| = (k-1)!(n-k)!$$

This is so because:

(1) The first $k - 1$ elements must be chosen from S in any of the $(k - 1)!$ possible orders.

(2) The remaining $n - k$ elements must be from $N \setminus (S \cup \{i\})$.

So one concludes

$$\pi_S^{Sh} = \frac{(k - 1)!(n - k)!}{n!} = \frac{(|S|!(n - |S| - 1)!}{n!}$$

and obtains another explicit formula for the SHAPLEY VALUE:

$$\Phi_i^{Sh}(v) = \sum_{S \subseteq N \setminus \{i\}} \partial_i v(S) \pi_S^{Sh} = \sum_{S \subseteq N \setminus \{i\}} \partial_i v(S) \frac{(|S|!(n - |S| - 1)!}{n!}$$

(66)

Ex. 8.23. Consider a voting/threshold game (*cf.* Section 3.4) with four players of weights $w_1 = 3, w_2 = 2, w_3 = 2, w_4 = 1$. Compute the BANZHAF and the SHAPLEY values for each of the players for the threshold $w = 4$.

9. Boltzmann values

The probabilistic analysis of the previous section shows that the value assessment concepts of the BANZHAF power index and the SHAPLEY value, for example, implicitly assume that players just join — but never leave — an existing coalition in a cooperative game (N, v).

In contrast, the model of the present section assumes an underlying probability distribution π on the set 2^N of *all* coalitions of N and assigns to player $i \in N$ its expected marginal value

$$E_i(v, \pi) = \sum_{S \subseteq N} \partial_i v(S) \pi_S.$$

Ex. 8.24. Let π be the uniform distribution on $\mathcal{N}$:

$$\pi_S = \frac{1}{|\mathcal{N}|} \quad \text{for all } S \in \mathcal{N}.$$

In view of

$$\sum_{S \subseteq N} \partial_i v(S) = \sum_{i \in S} v(S) - v(S \setminus i)) + \sum_{i \notin S} (v(S) - v(S \cup i))$$

$$= \sum_{T \subseteq N \setminus i} (v(T \cup i) - v(T)) + \sum_{T \subseteq N \setminus i} (v(T) - v(T \cup i))$$

$$= 0,$$

one has

$$E_i(v, \pi) = \sum_{S \in \mathcal{N}} \partial_i v(S) \pi_S = \frac{1}{|\mathcal{N}|} \sum_{S \in \mathcal{N}} \partial_i v(S) = 0.$$

So the expected marginal value of any particular player is zero, if all coalitions are equally likely.

We further do allow π to depend on the particular characteristic function v under consideration. It follows that the functional $v \mapsto E_i(v, \pi)$ is not guaranteed to be linear.

The idea of the BOLTZMANN value is based on the fact that one expects the characteristic value

$$\mu = E(v, \pi) = \sum_{S \in \mathcal{N}} v(S) \pi_S$$

if the players agree on a coalition $S \in \mathcal{N}$ with probability π_S. So we may associate with μ its BOLTZMANN temperature T and define the corresponding BOLTZMANN *values*

$$E_i^T(v) = \frac{1}{Z_T} \sum_{S \in \mathcal{N}} \partial_i v(S) e^{v(S)/T} \quad \text{(with } Z_T = \sum_{S \in \mathcal{N}} e^{v(S)/T}) \quad (67)$$

for the players i in the cooperative game (N, v) with expected characteristic value μ.

10. Coalition formation

In the cooperative game (N, v) it may not be clear in advance which coalition $S \subseteq N$ will form to play the game and produce the value $v(S)$. The idea behind the theory *coalition formation* is the viewpoint of a dynamic process in which players join and part in discrete time steps until a final coalition is likely to have emerged.

If this coalition formation process has BOLTZMANN temperature $T > 0$, the METROPOLIS process suggests the model where, at the moment of a current coalition $S \subseteq N$, a randomly determined player $i \in N$ considers to enact a possible transition

$$S \to S\Delta\{i\} = \begin{cases} S \setminus \{i\} & \text{if } i \in S \\ S \cup \{i\} & \text{if } i \notin S \end{cases}$$

by either becoming inactive or active. The transition is made with probability

$$\alpha_i(S) = \min\{1, e^{\partial_i v(S)/T}\}.$$

In this case, the coalition formation process converges to the corresponding BOLTZMANN distribution on the family $\mathcal{N}$ of coalitions in the limit.

Cost games. If (N, c) is a cost game, a player's gain from the transition $S \to S\Delta\{i\}$ is the negative marginal cost

$$-\partial_i c(S) = c(S) - c(S\Delta\{i\}).$$

So the transition $S \to S\Delta\{i\}$ in the METROPOLIS process would be enacted with probability

$$\alpha'_i(S) = \min\{1, e^{-\partial_i c(S)/T}\}.$$

10.1. Individual greediness and public welfare

Let us assume that N is a society whose common welfare is expressed by the potential v on the family $\mathcal{N}$ of all possible coalitions: If the members of N decide to join in a coalition $S \subseteq N$, then the value $v(S)$ will be produced.

If all members of N act purely greedily, an $i \in N$ has an incentive to change its decision with respect to the current coalition S depending on its marginal value $\partial_i v(S)$ being positive or negative. This behavior, however, will not guarantee a high public welfare.

The METROPOLIS process suggests that the public welfare can be steered if an incentive is provided such that i enacts a move $S \to S\Delta\{i\}$ (*i.e.*, changes its decision) with a non-zero probability

$$\alpha_i^T(S) = e^{\partial_i v(S)/T} \quad \text{(even) if } \partial_i v(S) < 0.$$

If the control parameter $T > 0$ is sufficiently small, the behavior of an $i \in N$ is "almost purely greedy" in the sense

$$T \to 0 \quad \Longrightarrow \quad \alpha_i^T(S) \to 0 \quad \text{if } \partial_i v(S) < 0.$$

Moreover, a small temperature $T > 0$ in the coalition formation process allows us to expect a high public welfare.

10.2. *Equilibria in cooperative games*

In many cooperative games, the grand coalition offers an obvious equilibrium if the players' utilities are assessed by their marginal values:

LEMMA 8.5. *Let (N, v) be a cooperative game. Then the two statements are equivalent:*

(1) *N is a gain equilibrium with respect to the individual utility functions $u_i(S) = \partial_i v(S)$.*

(2) *$v(N) \geq v(N \setminus i)$ for all $i \in N$.*

In general, we may view (N, v) as an n-person matrix game with individual utilities

$$u_i(S) = \partial_i v(S) = v(S\Delta i) - v(S).$$

Hence we know from NASH's Theorem 6.1 that the randomization of (N, v) admits an equilibrium.

REMARK 8.15. The randomization of (N, v) means that each $i \in N$ selects a probability $0 \leq w_i \leq 1$ for the probability to become active.

The coalition S is thus formed with probability

$$w(S) = \prod_{i \in S} w_i \prod_{j \notin S} (1 - w_j).$$

The expected value of v is thus

$$E(v, w) = \sum_{S \subseteq N} v(S)w(S).$$

The randomization of (N, v) essentially is a fuzzy game (see Ex. 6.2) with potential function

$$\overline{v}(v) = E(v, w) \quad (w \in [0, 1]^N).$$

Observe, in contrast to the above:

> • The BOLTZMANN coalition formation model does *not* admit coalition equilibria at temperature $T \neq 0$, unless v is constant,

but implies high public welfare if the temperature is small.

Many value concepts (like SHAPLEY and BANZHAF, for example), are based on marginal gains with respect to having *joined* a coalition as fundamental criteria for the individual utility assessment of a player.

So let us consider the cooperative game (N, v) and take into account that the game will eventually split N into a group $S \subseteq N$ and the complementary group $S^c = N \setminus S$. Suppose a player $i \in N$ evaluates its utility relative to the partition (S, S^c) of N by

$$v_i(S) = v_i(S^c) = \begin{cases} v(S) - v(S \setminus i) & \text{if } i \in S \\ v(S^c) - v(S^c \setminus i) & \text{if } i \in S^c. \end{cases}$$

Ex. 8.25. Assume that (N, v) is a supermodular game. Then one has for all players $i \neq j$,

$$v_i(N) = v(N) - v(N \setminus i) \geq v(N \setminus j) - v((N \setminus j) \setminus i) = v_i(N \setminus j)$$
$$v_i(N) = v(N) - v(N \setminus i) \geq v(\{i\}) - v(\emptyset) = v_i(N \setminus i).$$

Consequently, the grand coalition N represents a gain equilibrium relative to the utilities v_i.

Ex. 8.26. Assume that (N, c) is a zero-normalized submodular game and that the players i have the utilities

$$c_i(S) = \begin{cases} c(S) - c(S \setminus i) & \text{if } i \in S \\ c(S^c) - c(S^c \setminus i) & \text{if } i \in S^c. \end{cases}$$

Show: The grand coalition N is a cost equilibrium relative to the utilities c_i.

Chapter 9

Interaction Systems and Quantum Models

This final chapter investigates game-theoretic systems making use of the algebra of complex numbers. Not only cooperation models are generalized, but also interaction of pairs on elements of a set X finds an appropriate setting. The states are naturally represented as hermitian matrices with complex coefficients. This representation allows one to carry out standard spectral analysis for interaction systems and provides a link to the corresponding mathematical model of quantum systems in physics.

While the analysis could be extended to general HILBERT spaces, X is assumed to be finite to keep the discussion straigthforward.[a]

It is historically perhaps surprising that JOHN VON NEUMANN, who laid out the mathematical foundations of quantum theory,[b] did not build game theory on the same mathematics in his work with OSKAR MORGENSTERN.

---

[a]See also FAIGLE and GRABISCH [12]
[b]VON NEUMANN [33]

1. Algebraic preliminaries

Since matrix algebra is the main tool in our analysis, we review some more fundamental notions from linear algebra (and Chapter 2). Further details and proofs can be found in any decent book on linear algebra.[1]

---

[1]*e.g.*, NERING [31].

Where $X = \{x_1, \ldots, x_m\}$ and $Y = \{y_1, \ldots, y_n\}$ are two finite index sets, recall that $\mathbb{R}^{X \times Y}$ denotes the real vector space of all matrices A with rows indexed by X, columns indexed by Y, and coefficients $A_{xy} \in \mathbb{R}$.

The *transpose* of $A \in \mathbb{R}^{X \times X}$ is the matrix $A^T \in \mathbb{R}^{Y \times X}$ with the coefficients $A^T_{xy} = A_{xy}$. The map $A \mapsto A^T$ establishes an isomorphism between the vector spaces $\mathbb{R}^{X \times Y}$ and $\mathbb{R}^{Y \times X}$.

Viewing $A \in \mathbb{R}^{X \times Y}$ and $B \in \mathbb{R}^{Y \times X}$ as mn-dimensional parameter vectors, we have the usual euclidian inner product as

$$\langle A|B \rangle = \sum_{(x,y) \in X \times Y} A_{xy} B_{xy} = \mathrm{tr}\,(B^T A),$$

where tr C denotes the trace of a matrix C. In the case $\langle A|B \rangle = 0$, A and B are said to be *orthogonal*. The associated euclidian norm is

$$\|A\| = \sqrt{\langle A|A^T \rangle} = \sqrt{\sum_{(x,y) \in X \times Y} |A_{xy}|^2}.$$

We think of a vector $v \in \mathbb{R}^X$ typically as a column vector. v^T is the row vector with the same coordinates $v^T_x = v_x$. Be aware of the difference between the two matrix products:

$$v^T v = \sum_{x \in X} |v_x|^2 = \|v\|^2$$

$$vv^T = \begin{bmatrix} v_{x_1} v_{x_1} & v_{x_1} v_{x_2} & \cdots & v_{x_1} v_{x_m} \\ v_{x_2} v_{x_1} & v_{x_2} v_{x_2} & \cdots & v_{x_1} v_{x_m} \\ \vdots & \vdots & \ddots & \vdots \\ v_{x_m} v_{x_1} & v_{x_m} v_{x_2} & \cdots & v_{x_m} v_{x_m} \end{bmatrix}.$$

1.1. *Symmetry decomposition*

Assuming identical index sets

$$X = Y = \{x_1, \ldots, x_n\}$$

a matrix $A \in \mathbb{R}^{X \times X}$ is *symmetric* if $A^T = A$. In the case $A^T = -A$, the matrix A is *skew-symmetric*. With an arbitrary matrix

$A \in \mathbb{R}^{X \times X}$, we associate the matrices

$$A^+ = \frac{1}{2}(A + A^T) \quad \text{and} \quad A^- = \frac{1}{2}(A - A^T) = A - A^+.$$

Notice that A^+ is symmetric and A^- is skew-symmetric. The *symmetry decomposition* of A is the representation

$$\boxed{A = A^+ + A^-} \tag{68}$$

The matrix A allows exactly one decomposition into a symmetric and a skew-symmetric matrix (see Ex. 9.1). So the symmetry decomposition is unique.

Ex. 9.1. Let $A, B, C \in \mathbb{R}^{X \times X}$ be such that $A = B + C$. Show that the two statements are equivalent:

(1) B is symmetric and C is skew-symmetric.
(2) $B = A^+$ and $C = A^-$.

Notice that symmetric and skew-symmetric matrices are necessarily pairwise orthogonal (see Ex. 9.2).

Ex. 9.2. Let A be a symmetric and B a skew-symmetric matrix. Show:

$$\langle A|B \rangle = 0 \quad \text{and} \quad \|A + B\|^2 = \|A\|^2 + \|B\|^2.$$

2. Complex matrices

In physics and engineering, complex numbers offer a convenient means to represent orthogonal structures. Applying this idea to the symmetry decomposition, one arrives at so-called *hermitian matrices*.

Inner products. Recall that a *complex number* $z \in \mathbb{C}$ is an expression of the form $z = a + ib$ where a and b are real numbers and i a special "new" number, the *imaginary unit*, with the property

$i^2 = -1$. The squared absolute value of the complex number $z = a + ib$ is

$$|z|^2 = a^2 + b^2 = (a - ib)(a + ib) = \overline{z}z,$$

with $\overline{z} = a - ib$ being the *conjugate* of z. More generally, we define the *hermitian product* of two complex numbers $u, v \in \mathbb{C}$ as the complex number

$$\langle u|v \rangle = \overline{v}u. \tag{69}$$

The *(hermitian) inner product* of two vectors $u, v \in \mathbb{C}^X$ with components u_x and v_x is the complex number

$$\langle u|v \rangle = \sum_{x \in X} \langle u_x|v_x \rangle.$$

The *length* (or *norm*) of a vector $u = a + ib \in \mathbb{C}^X$ (with $a, b \in \mathbb{R}^X$) is

$$\|u\| = \sqrt{\sum_{x \in X} \langle u_x|u_x \rangle} = \sqrt{\sum_{x \in X} |u_x|^2} = \sqrt{\sum_{x \in X} |a_x|^2 + |b_x|^2}.$$

Conjugates and adjoints. The *conjugate* of a vector $v \in \mathbb{C}^X$ is the vector $\overline{v} \in \mathbb{C}^X$ with the conjugated components $\overline{v}_x$. The vector $v^* = \overline{v}^T$ is the *adjoint* of v. With this notation, the inner product of the column vectors $u, v \in \mathbb{C}^X$ is

$$\langle u|v \rangle = \sum_{x \in X} u_x v_x^* = v^* u,$$

where we think of the 1×1 matrix $v^* u$ just as a complex number. Accordingly, the adjoint of the matrix $C \in \mathbb{C}^{X \times Y}$ is the matrix

$$C^* = \overline{C}^T \in \mathbb{C}^{Y \times X}.$$

Ex. 9.3 (Trace). The inner product of the matrices $U, V \in \mathbb{C}^{X \times Y}$ is

$$\langle U|V \rangle = \text{tr} \, (V^* U).$$

The matrix $C \in \mathbb{C}^{X \times X}$ is *selfadjoint* if it equals its adjoint, *i.e.*, if

$$C = C^* = \overline{C}^T.$$

Ex. 9.4. Let $v \in \mathbb{C}^X$ be a column vector. Then $vv^* \in \mathbb{C}^{X \times X}$ is a selfadjoint matrix of norm $\|vv^*\| = \|v\|^2$.

2.1. Spectral decomposition

If (and only if) the matrix $C \in \mathbb{C}^{X \times X}$ has real coefficients,

$$\overline{C} = C$$

holds and the notion "selfadjoint" boils down to "symmetric". It is well-known that real symmetric matrices can be diagonalized. With the same technique, one can extend the diagonalization to general selfadjoint matrices:

THEOREM 9.1 (Spectral Theorem). *For a matrix $C \in \mathbb{C}^{X \times X}$ the two statements are equivalent:*

(1) $C = C^*$.
(2) $\mathbb{C}^X$ *admits a unitary basis* $U = \{U_x \mid x \in X\}$ *of eigenvectors* U_x *of C with real eigenvalues* λ_x.

Unitary means for the basis U that the vectors U_x have unit norm and are pairwise orthogonal in the sense

$$\langle U_x | U_y \rangle = \begin{cases} 1 & \text{if } x = y \\ 0 & \text{if } x \neq y. \end{cases}$$

The scalar λ_x is the *eigenvalue* of the *eigenvector* U_x of C if

$$CU_x = \lambda_x U_x.$$

It follows from Theorem 9.1 (see Ex. 9.5) that a selfadjoint matrix C admits a *spectral*[2] *decomposition, i.e.*, a representation in the form

---

[2] The *spectrum* of a matrix is, by definition, its set of eigenvalues.

$$C = \sum_{x \in X} \lambda_x U_x U_x^*, \tag{70}$$

where the U_x are pairwise orthogonal eigenvectors of C with eigenvalues $\lambda_x \in \mathbb{R}$.

Ex. 9.5. Let $U = \{U_x \mid x \in X\}$ be a unitary basis of $\mathbb{C}^X$ together with a set $\Lambda = \{\lambda_x \mid x \in X\}$ a set of arbitrary complex scalars. Show:

(1) The U_x are eigenvectors with eigenvalues λ_x of the matrix

$$C = \sum_{x \in X} \lambda_x U_x U_x^*.$$

(2) C is selfadjoint if and only if all the λ_x are real numbers.

The spectral decomposition shows:

The selfadjoint matrices C in $\mathbb{C}^{X \times X}$ are precisely the linear combinations of matrices of type

$$C = \sum_{x \in X} \lambda_x U_x U_x^*,$$

where the U_x are (column) vectors in $\mathbb{C}^X$ and the λ_x are real numbers.

Spectral unity decomposition. As an illustration, consider a unitary matrix $U \in \mathbb{C}^{X \times X}$, *i.e.*, a matrix with pairwise orthogonal column vectors U_x of norm $\|U\|_x = 1$, which means that the identity matrix I has the representation

$$I = UU^* = U^*U.$$

The eigenvalues of I have all value $\lambda_x = 1$. Relative to U, the matrix I has the spectral decomposition

$$I = \sum_{x \in X} U_x U_x^*. \tag{71}$$

For any vector $v \in \mathbb{C}^X$ with norm $\|v\| = 1$, we therefore find

$$1 = \langle v|v \rangle = v^* I v = \sum_{x \in X} v^* U_x U_x^* v$$

$$= \sum_{x \in X} \overline{\langle v|U_x \rangle} \langle v|U_x \rangle = \sum_{x \in X} |\langle v|U_x \rangle|^2.$$

It follows that the (squared) absolute values

$$p_x^v = |\langle v|U_x \rangle|^2 \quad (x \in X)$$

yield the components of a probability distribution p^v on the set X. More generally, if the selfadjoint matrix C with eigenvalues ρ_x has the form

$$C = \sum_{x \in X} \rho_x U_x U_x^*,$$

then we have for any $v \in \mathbb{C}^X$,

$$\langle v|Cv \rangle = v^* C v = \sum_{x \in X} \rho_x |\langle v|U_x \rangle|^2 = \sum_{x \in X} \rho_x p_x^v. \tag{72}$$

In other words:

> The inner product $\langle v|Cv \rangle$ of the vectors v and Cv is the expected value of the eigenvalues ρ_x of C with respect to the probability distribution p^v on X.

Ex. 9.6 (Standard unity decomposition). The unit vectors $e_x \in \mathbb{C}^X$ yield the standard unity decomposition

$$I = \sum_{x \in X} e_x e_x^*.$$

Accordingly, a vector $v \in \mathbb{C}^X$ of length $\|v\| = 1$ with the components v_x implies the standard probability distribution on X with the components

$$p_x^v = |\langle v|e_x \rangle|^2 = |v_x|^2.$$

2.2. Hermitian representation

Coming back to real matrices in the context of symmetry decompositions, let us associate with a real matrix $A \in \mathbb{R}^{X \times X}$ the complex matrix

$$\hat{A} = A^+ + iA^-.$$

$\hat{A}$ is a *hermitian*[3] *matrix*. The *hermitian map* $A \mapsto \hat{A}$ establishes an isomorphism between the vector space $\mathbb{R}^{X \times X}$ and the vector space

$$\mathbb{H}_X = \{\hat{A} \mid A \in \mathbb{R}^{X \times X}\}$$

with the set $\mathbb{R}$ as field of scalars.[4] The import in our context is the fundamental observation that the selfadjoint matrices are precisely the hermitian matrices:

LEMMA 9.1. *Let* $C \in \mathbb{C}^{X \times X}$ *be an arbitrary complex matrix. Then*

$$C \in \mathbb{H}_X \quad \Longleftrightarrow \quad C = C^*$$

Proof. Assume $C = A + iB$ with $A, B \in \mathbb{R}^{X \times}$ and hence

$$C^* = A^T - iB^T$$

So $C = C^*$ means symmetry $A = A^T$ and skew-symmetry $B = -B^T$. Consequently, one has $\hat{A} = A$ and $\hat{B} = iB$, which yields

$$C = A + iB = \hat{A} + \hat{B} \in \mathbb{H}_X.$$

The converse is seen as easily. $\square$

---

[3] C. HERMITE (1822–1901).

[4] $\mathbb{H}_X$ is not a complex vector space: The product zC of a hermitian matrix C with a complex scalar z is not necessarily hermitian.

The remarkable property of the hermitian representation is:

> • While a real matrix $A \in \mathbb{R}^{X \times X}$ does not necessarily admit a spectral decomposition with real eigenvalues, its hermitian representation $\hat{A}$ is always guaranteed to have one.

Ex. 9.7 (HILBERT space). Let $A, B \in \mathbb{R}^{X \times X}$ be arbitrary real matrices. Show:

$$\langle A | B \rangle = \langle \hat{A} | \hat{B} \rangle,$$

i.e., inner products (and hence norms) are preserved under the hermitian representation. This means that $\mathbb{R}^{X \times X}$ and $\mathbb{H}_X$ are not only isomorphic as real vector spaces but also as (real) HILBERT spaces.

3. Interaction systems

Let us assume that elements $x, y \in X$ can *interact* with a certain *interaction strength*, measured by a real number a_{xy}. We denote this interaction symbolically as $a_{xy}\varepsilon_{xy}$. Graphically, one may equally well think of a weighted (directed) edge in an *interaction graph* with X as its set of nodes:

$$a_{xy}\varepsilon_{xy} \quad :: \quad (\text{x}) \xrightarrow{a_{xy}} (\text{y})$$

An *interaction instance* is a weighted superposition of interactions:

$$\varepsilon = \sum_{x,y \in X} a_{ax}\varepsilon_{xy}.$$

We record the interaction instance ε in the *interaction matrix* $A \in \mathbb{R}^{X \times X}$ with the *interaction coefficients* $A_{xy} = a_{xy}$. The interaction is *symmetric* if $A^T = A$ and *skew-symmetric* if $A^T = -A$.

Conversely, each matrix $A \in \mathbb{R}^{X \times X}$ corresponds to some interaction instance

$$\varepsilon = \sum_{x,y \in X} A_{xy} \varepsilon_{xy}.$$

So we may think of $\mathbb{R}^{X \times X}$ as the *interaction space* relative to the set X. Moreover, the symmetry decomposition

$$A = A^+ + A^-$$

shows:

Every interaction instance ε on X is the superposition

$$\varepsilon = \varepsilon^+ + \varepsilon^-$$

of a symmetric interaction instance ε^+ and a skew-symmetric interaction instance ε^-. Moreover, ε^+ and ε^- are uniquely determined.

Binary interaction. In the binary case $|X| = 2$, interaction matrices are (2×2)-matrices. Consider, for example, the matrices

$$I = \begin{bmatrix} 1 & 0 \\ 0 & 1 \end{bmatrix} \quad \text{and} \quad Im = \begin{bmatrix} 0 & 1 \\ -1 & 0 \end{bmatrix}.$$

I is symmetric and Im is skew-symmetric. Superposition of these two matrices with real scalars $a, b \in \mathbb{R}$ yields the interaction matrix

$$Z = aI + bIm = \begin{bmatrix} a & b \\ -b & a \end{bmatrix},$$

which is the matrix representation of the complex number $z = a + ib$.

Note that the binary interaction space is 4-dimensional, however. So the complex numbers only describe a 2-dimensional subclass of binary interactions.

3.1. *Interaction states*

The *norm* of an interaction state ε with interaction matrix A is the norm of the associated interaction matrix:

$$\|\varepsilon\| = \|A\|.$$

So $\|\varepsilon\| \neq 0$ means that at least two members $s, t \in X$ interact with strength $A_{st} \neq 0$ and that the numbers

$$p_{xy} = \frac{|A_{xy}|^2}{\|A\|^2} \quad ((x, y) \in X \times X)$$

yield a probability distribution on the set of all possibly interacting pairs and offer a probabilistic perspective on ε:

> - *A pair (x, y) of members of X is interacting nontrivially with probability p_{xy}.*

Clearly, scaling ε to $\lambda\varepsilon$ with a scalar $\lambda \neq 0$, would result in the same probability distribution on $X \times X$. From the probabilistic point of view, it therefore suffices to consider interaction instances ε with norm $\|\varepsilon\| = 1$.

We thus define:

> *The interaction system on X is the system $\mathfrak{I}(X)$ with the set of states*
>
> $$\mathfrak{I}_X = \{\varepsilon \mid \varepsilon \text{ is an interaction instance of } X \text{ of norm } \|\varepsilon\| = 1\}$$

In terms of the matrix representation of states, we have

$$\mathfrak{I}_X \quad \longleftrightarrow \quad \mathcal{S}_X = \{A \in \mathbb{R}^{X \times X} \mid \|A\| = 1\}.$$

3.2. *Interaction potentials*

A potential $F : X \times X \to \mathbb{R}$ defines a matrix with coefficients $F_{xy} = F(x, y)$ and thus a scalar-valued linear functional

$$A \mapsto \langle F|A \rangle = \sum_{x,y \in X} F_{xy} A_{xy}$$

on the vector space $\mathbb{R}^{X \times X}$. Conversely, every linear functional f on $\mathbb{R}^{X \times X}$ is of the form

$$f(A) = \sum_{x,y \in X} F_{xy} A_{xy} = \langle F | A \rangle$$

with uniquely determined coefficients $F_{xy} \in \mathbb{R}$. So potentials and linear functionals correspond to each other.

On the other hand, the potential F defines a linear operator $A \mapsto F \bullet A$ on the space $\mathbb{R}^{X \times X}$, where the matrix $F \bullet A$ is the HADAMARD *product* of F and A with the coefficients

$$(F \bullet A)_{xy} = F_{xy} A_{xy} \quad \text{for all } x, y \in X.$$

With this understanding, one has

$$\langle F | A \rangle = \sum_{x,y \in X} (F \bullet A)_{xy}.$$

Moreover, one computes

$$\boxed{\langle A | F \bullet A \rangle = \sum_{x,y \in X} A_{xy} (F \bullet A)_{xy} = \sum_{x,y \in X} F_{xy} |A_{xy}|^2.} \qquad (73)$$

Summarizing, one finds:

If $A \in \mathcal{S}_X$ (*i.e.*, if A represents an interaction state $\varepsilon \in \mathfrak{I}_X$), the parameters $p_{xy}^A = |A_{xy}|^2$ define a probability distribution on $X \times X$. The expected value of the potential F in this state ε is

$$\mu^\varepsilon(F) = \sum_{x,y \in X} F_{xy} p_{xy}^A = \langle A | F \bullet A \rangle.$$

3.3. *Interaction in cooperative games*

The interaction model offers a considerably wider context for the analysis of cooperation.

For an illustration, consider a cooperative TU-game $\Gamma = (N, v)$ with collection $\mathcal{N}$ of coalitions. v is a potential on $\mathcal{N}$ but not on the set $\mathcal{N} \times \mathcal{N}$ of possibly pairwise interacting coalitions. However, there is a straightforward extension of v to $\mathcal{N} \times \mathcal{N}$:

$$v(S,T) = \begin{cases} v(S) & \text{if } S = T \\ 0 & \text{if } S \neq T. \end{cases}$$

Relative to a state $\sigma \in \mathfrak{I}_{\mathcal{N} \times \mathcal{N}}$ with interaction matrix A, the expected value of v is

$$v(\sigma) = \sum_{S \in \mathcal{N}} v(S)|A_{SS}|^2.$$

In the special case of a state σ_S where the coalition S interacts with itself with certainty (and hence no proper interaction among other coalitions takes place), we have

$$v(\sigma_S) = v(S)$$

which is exactly the potential value of the coalition S in the classical interpretation of Γ.

Generalized cooperative games. A more comprehensive model for the study of cooperation among players would be structures of the type

$$\Gamma = (N, \mathcal{N}, v),$$

where v is a potential on $\mathcal{N} \times \mathcal{N}$ (rather than just $\mathcal{N}$). This point of view suggests to study game-theoretic cooperation within the context of interaction.

3.4. *Interaction in infinite sets*

Much of the current interaction analysis remains valid for infinite sets with some modifications.

For example, we admit as descriptions of interaction states only those matrices $A \in \mathbb{R}^{X \times X}$ with the property

(H1) $\text{supp}(A) = \{(x,y) \in X \times X \mid A_{xy} \neq 0\}$ is finite or countably infinite.

(H2) $\|A\|^2 = \displaystyle\sum_{x,y \in X} |A_{xy}|^2 = 1.$

If the conditions (H1) and (H2) are met, we factually represent interaction states in HILBERT spaces. To keep things simple, however, we retain the finiteness property of the agent set X in the current text and refer the interested reader to the literature[5] for further details.

4. Quantum systems

Without going into the physics of quantum mechanics, let us quickly sketch the basic mathematical model and then look at the relationship with the interaction model. In this context, we think of an *observable* as a mechanism α that can be applied to a system $\mathfrak{S}$,

$$\boxed{\sigma} \rightsquigarrow \boxed{\alpha} \longrightarrow \alpha(\sigma)$$

with the interpretation:

> • If $\mathfrak{S}$ is in the state σ, then α is expected to produce a measurement result $\alpha(\sigma)$.

4.1. *The quantum model*

There are two views on a *quantum system* $\mathfrak{Q}_X$ relative to a set X. They are dual to each other (reversing the roles of states and observables) but mathematically equivalent.

The SCHRÖDINGER picture. In the so-called SCHRÖDINGER[6] *picture*, the states of $\mathfrak{Q}_X$ are presented as the elements of the set

$$\mathcal{W}_X = \{v \in \mathbb{C}^X \mid \|v\| = 1\}$$

of complex vectors of norm 1. An observable α corresponds to a (selfadjoint) matrix $A \in \mathbb{H}_X$ and produces the real number

$$\alpha(v) = \langle v|Av \rangle = v^*A^*v = v^*Av$$

---

[5] *e.g.*, HALMOS [23] or WEIDMANN [46].
[6] E. SCHRÖDINGER (1887–1961).

when $\mathfrak{Q}_X$ is in the state $v \in \mathcal{W}$. Recall from the discussion of the spectral decomposition in Section 2.1 that $\alpha(v)$ is the expected value of the eigenvalues ρ_x of A relative to the probabilities

$$p_x^{A,v} = |\langle v | U_x \rangle|^2 \quad (x \in X),$$

where the vectors $U_x \in \mathcal{W}$ constitute a vector space basis of corresponding unitary eigenvectors of A.

An interpretation of the probabilities $p^{A,v}$ could be as follows:

$\mathfrak{Q}_X$ is a stochastic system that shows the element $x \in X$ with probability $p_x^{A,v}$ if it is observed under A in the state v:

$$v \rightsquigarrow \boxed{A} \longrightarrow x.$$

The elements of X are weighted with the eigenvalues of A. The measurement value is the expected weight of the x produced under A.

Ex. 9.8. The identity matrix $I \in \mathbb{C}^{X \times X}$ yields the standard probabilities[7]

$$p_x^{I,v} = |v_x|^2 \quad (x \in X).$$

The HEISENBERG picture. In the HEISENBERG[8] *picture* of $\mathfrak{Q}_X$, the selfadjoint matrices $A \in \mathbb{H}_X$ take over the role of states while the vectors $v \in \mathcal{W}$ induce measurement results. The HEISENBERG picture is dual[9] to the SCHRÖDINGER picture. In both pictures, the expected values

$$\langle v | Av \rangle \quad (v \in \mathcal{W}_X, A \in \mathbb{H}_X)$$

are thought to be the numbers resulting from measurements on $\mathfrak{Q}_X$.

---

[7] *cf.* Ex. 9.6.
[8] W. HEISENBERG (1901–1976).
[9] In the sense of Section 2.1.

The HEISENBERG picture sees an element $x \in X$ according to the scheme

$$A \quad \longrightarrow \quad \boxed{v} \quad \rightsquigarrow \quad x$$

with probability $p_x^{A,v}$.

Densities and wave functions. The difference of the two pictures lies in the interpretation of the probability distribution $p^{A,v}$ on the index set X relative to $A \in \mathbb{H}_X$ and $v \in \mathcal{W}_X$.

In the HEISENBERG picture, $p^{A,v}$ is imagined to be implied by a possibly varying A relative to a fixed state vector v. Therefore, the elements $A \in \mathbb{H}_X$ are also known as *density matrices*.

In the SCHRÖDINGER picture, the matrix A is considered to be fixed while the state vector $v = v(t)$ may vary in time t. $v(t)$ is called a *wave function*.

4.2. *Evolutions of quantum systems*

A quantum evolution

$$\Phi = \Phi(M, v, A)$$

in (discrete) time t depends on a matrix-valued function $t \mapsto M_t$, a state vector $v \in \mathcal{W}$, and a density $A \in \mathbb{H}_X$. The evolution Φ produces real observation values

$$\varphi_t = v^*(M_t^* A M_t)v \quad (t = 0, 1, 2, \ldots). \tag{74}$$

Notice that the matrices $A_t = M_t^* A M_t$ are selfadjoint. So the evolution Φ can be seen as an evolution of density matrices, which is in accordance with the HEISENBERG picture.

If $v(t) = M_t v \in \mathcal{W}$ holds for all t, the evolution Φ can also be interpreted in the SCHRÖDINGER picture as an evolution of state vectors:

$$\varphi_t = (M_t v)^* A (M_t v) \quad (t = 0, 1, 2, \ldots). \tag{75}$$

REMARK 9.1. The standard model of quantum mechanics assumes that evolutions satisfy the condition $M_t v \in \mathcal{W}$ at any time t, so that the HEISENBERG and the SCHRÖDINGER pictures are equivalent.

Markov coalition formation. Let $\mathcal{N}$ be the collection of coalitions of the set N. The classical view on coalition formation sees the probability distributions p on $\mathcal{N}$ as the possible states of the formation process and the process itself as a MARKOV[10] chain. For example, the METROPOLIS process[11] is an example of a MARKOV chain.

To formalize this model, let $\mathfrak{P} = \mathfrak{P}(N)$ be the set of all probability distributions on $\mathcal{N}$. A MARKOV *operator* is a linear map

$$\mu : \mathbb{N}_0 \to \mathbb{R}^{\mathcal{N} \times \mathcal{N}} \quad \text{such that } \mu_t p \in \mathfrak{P} \text{ holds for all } p \in \mathfrak{P}.$$

μ defines for every initial state $p^{(0)} \in \mathfrak{P}$ a so-called MARKOV *chain* of probability distributions

$$\mathcal{M}(p^{(0)}) = \{\mu^t(p^{(0)}) \mid t = 0, 1, \ldots\}.$$

Define now $P_t \in \mathbb{R}^{\mathcal{N} \times \mathcal{N}}$ as the diagonal matrix with $p^{(t)} = \mu^t(p^{(0)})$ as its diagonal coefficient vector. P_t is a real symmetric matrix and therefore a density in particular.

Any $v \in \mathcal{W}$ gives rise to a quantum evolution with observed values

$$\pi_t = v^* P_t v \quad (t = 0, 1, \ldots, n).$$

For example, if $e_S \in \mathbb{R}^{\mathcal{N}}$ is the unit vector that corresponds to the coalition $S \in \mathcal{N}$, one has

$$\pi_t^{(S)} = e_S^* P_t e_S = (P_t)_{SS} = p_S^{(t)}$$

with the usual interpretation:

---

[10] A.A. MARKOV (1856–1922).
[11] *cf.* Chapter 2.

- If the coalition formation proceeds according to the MARKOV chain $\mathcal{M}(p^{(0)})$, then an inspection at time t will find the coalition S to be active with probability $\pi_t^{(S)} = p_S^{(t)}$.

4.3. The quantum perspective on interaction

Recalling the vector space isomorphism *via* the hermitian representation

$$A \in \mathbb{R}^{X \times X} \longleftrightarrow \hat{A} = A^+ \mathrm{i} A^- \in H_X,$$

we may think of interaction states as manifestations of SCHRÖDINGER states of the quantum system $\mathfrak{Q}_{X \times X}$,

$$\mathcal{S}_X = \{ A \in \mathbb{R}^{X \times X} \mid \|A\| = 1 \} \leftrightarrow \mathcal{W}_{X \times X} = \{ \hat{A} \in \mathbb{H}_X \mid \|\hat{A}\| = 1 \},$$

or as normed representatives of HEISENBERG densities relative to the quantum system $\mathfrak{Q}_X$.

Principal components. An interaction instance A on X has a hermitian spectral decomposition

$$\hat{A} = \sum_{x \in X} \lambda_x U_x U_x^* = \sum_{x \in X} \lambda_x \hat{A}_x$$

where the matrices $\hat{A}_x = U_x U_x^*$ are the *principal components* of $\hat{A}$. The corresponding interaction instances A_x are the principal components of A:

$$A = \sum_{x \in X} \lambda_x A_x.$$

Principal components V of interaction instances arise from SCHRÖDINGER states $v = a + \mathrm{i}b \in \mathcal{W}_X$ with $a, b \in \mathbb{R}^X$ in the following way. Setting

$$\hat{V} = vv^* = (a + \mathrm{i}b)(a - \mathrm{i}b)^T = aa^T - bb^T + \mathrm{i}(ba^T - ab^T),$$

one has $V^+ = aa^T + bb^T$ and $V^- = ba^T - ab^T$, and thus

$$V = V^+ + V^- = (aa^T + bb^T) + (ba^T - ab^T).$$

The principal interaction instance V has hence the underlying structure:

---

(0) Each $x \in X$ has a pair (a_x, b_x) of weights $a_x, b_x \in \mathbb{R}$.

(1) The symmetric interaction between two arbitrary elements $x, y \in X$ is

$$V_{xy}^+ = a_x a_y + b_x b_y.$$

(2) The skew-symmetric interaction between two arbitrary elements $x, y \in X$ is

$$V_{xy}^- = b_x a_y - a_x b_y.$$

4.4. *The quantum perspective on cooperation*

Let N be a (finite) set of players and family $\mathcal{N}$ of coalitions. From the quantum point view, a (SCHRÖDINGER) *state* of N is a complex vector $v \in \mathcal{W}_N$, which implies the probability distribution p^v with probabilities

$$p_i^v = |v_i|^2 \quad (i \in N)$$

on N. In the terminology of *fuzzy cooperation*, p^v describes a fuzzy coalition:

• Player $i \in N$ is *active* in state v with probability p_i^v.

Conversely, if $w \in \mathbb{R}^N$ is a non-zero fuzzy coalition with component probabilities $0 \le w_i \le 1$, the vector

$$\sqrt{w} = (\sqrt{w_i} \mid i \in N)$$

may be normalized to a SCHRÖDINGER state

$$v = \frac{\sqrt{w}}{\|\sqrt{w}\|} \quad \text{s.t.} \quad w_i = \|\sqrt{w}\| \cdot |v_i|^2 \quad \text{for all } i \in N.$$

In the same way, a vector $\mathcal{W_N}$ describes a SCHRÖDINGER state of interaction among the coalitions of N. It is particularly instructive to look at the interactions V of principal component type.

As we have seen in Section 4.3 above, V arises as follows:

---

(0) The interaction V on $\mathcal{N}$ is implied by two cooperative games $\Gamma_a = (N, a)$ and $\Gamma_b = (N, b)$.

(1) Two coalitions $S, T \in \mathcal{N}$ interact symmetrically *via*

$$V_{ST}^+ = a(S)a(T) + b(S)b(T).$$

(2) Two coalitions $S, T \in \mathcal{N}$ interact skew-symmetrically *via*

$$V_{ST}^- = b(S)a(T) - a(S)b(T).$$

---

5. Quantum games

A large part of the mathematical analysis of game-theoretic systems follows the guideline:

- *Represent the system in a mathematical structure, analyze the representation mathematically and re-interpret the result in the original game-theoretic setting.*

When one chooses a representation of the system in the same space as the ones usually employed for the representation of a quantum system, one automatically arrives at a "quantum game", *i.e.*, at a quantum-theoretic interpretation of a game-theoretic environment.

So we understand by a *quantum game* any game on a system $\mathfrak{S}$ whose states are represented as quantum states and leave it to the reader to review game theory in this more comprehensive context.

6. Final remarks

Why should one pass to complex numbers and the hermitian space $\mathbb{H}_X$ rather than the euclidian space $\mathbb{R}^{X \times X}$ if both spaces are isomorphic real Hilbert spaces?

The advantage lies in the algebraic structure of the field $\mathbb{C}$ of complex numbers, which yields the spectral decomposition (70), for example. It would be not impossible, but somewhat "unnatural" to translate this structural insight back into the environment $\mathbb{R}^{X \times X}$ without appeal to complex algebra.

Another advantage becomes apparent when one studies evolutions of systems over time. In the classical situation of real vector spaces, MARKOV chains are important models for system evolutions. It turns out that this model generalizes considerably when one passes to the context of HILBERT spaces.[12]

The game-theoretic ramifications of this approach are to a large extent unexplored at this point.

---

[12]FAIGLE and GIERZ [11].

Appendix

1. Basic facts from real analysis

(More details can be found in the standard literature.[1]) The *euclidian norm* (or *geometric length*) of a vector $x \in \mathbb{R}^n$ with components x_j, is

$$\|x\| = \sqrt{x_1^2 + \ldots + x_n^2}.$$

The *ball* with center x and radius r is the set

$$B_r(x) = \{y \in \mathbb{R}^n \mid \|x - y\| \leq r\}.$$

A subset $S \subseteq \mathbb{R}^n$ is *closed* if each convergent sequence of elements $x_k \in S$ has its the limit point x is also in S:

$$x_k \to x \quad \Longrightarrow \quad x \in S.$$

The set S is *open* if its complement $\mathbb{R}^n \setminus S$ is closed. The following statements are equivalent:

(O') S is open.
(O'') For each $x \in S$ there is some $r > 0$ such that $B_r(x) \subseteq S$.

The set S is *bounded* if $S \subseteq B_r(0)$ holds for some $r \geq 0$. S is said to be *compact* if S is bounded and closed.

---

[1] *e.g.*, RUDIN [39].

LEMMA A.2 (HEINE–BOREL). $S \subseteq \mathbb{R}^n$ *is compact if and only if*

(HB) *every family $\mathcal{O}$ of open sets $O \subseteq \mathbb{R}^n$ such that every $x \in S$ lies in at least one $O \in \mathcal{O}$, admits a finite number of sets $O_1, \ldots, O_\ell \in \mathcal{O}$ with the covering property*

- $S \subseteq O_1 \cup O_2 \cup \ldots \cup O_\ell$. $\qquad\qquad\Box$

It is important to note that compactness is preserved under forming direct products:

- If $X \subseteq \mathbb{R}^n$ and $Y \subseteq \mathbb{R}^m$ are compact sets, then $X \times Y \subseteq \mathbb{R}^{n+m}$ is compact.

Continuity. A function $f : S \to \mathbb{R}^m$ is *continuous* if for all convergent sequences of elements $x_k \in S$, the sequence of function values $f(x_k)$ converges to the value of the limit:

$$x_k \to x \implies f(x_k) \to f(x).$$

The following statements are equivalent:

(C′) $f : S \to \mathbb{R}^m$ is continuous.
(C″) For each open set $O \subseteq \mathbb{R}^m$, there exists an open set $O' \subseteq \mathbb{R}^n$ such that

$$f^{-1}(O) = \{x \in S \mid f(x) \in O\} = O' \subseteq S,$$

i.e., the inverse image $f^{-1}(O)$ is open *relative to S*.
(C″′) For each closed set $C \subseteq \mathbb{R}^m$, there exists a closed set $C' \subseteq \mathbb{R}^n$ such that

$$f^{-1}(C) = \{x \in S \mid f(x) \in C\} = C' \subseteq S,$$

i.e., the inverse image $f^{-1}(C)$ is closed *relative to S*.

LEMMA A.3 (Extreme values). *If the real-valued function $f : S \to \mathbb{R}$ is continuous on the nonempty compact set $S \subseteq \mathbb{R}^n$, then there exist elements $x_*, x^* \in S$ such that*

$$f(x_*) \leq f(x) \leq f(x^*) \quad \text{holds for all } x \in S. \qquad\Box$$

Differentiability. The function $f : S \to \mathbb{R}$ is *differentiable* on the open set $S \subseteq \mathbb{R}^n$ if for each $x \in S$ there is a (row) vector $\nabla f(x)$ such that for every $d \in \mathbb{R}^n$ of unit length $\|d\| = 1$, one has

$$\lim_{t \to 0} \frac{f(x + td) - f(x)}{t} = \lim_{t \to 0} \frac{\nabla f(x) d}{t} \quad (t \in \mathbb{R}).$$

$\nabla f(x)$ is the *gradient* of f. Its components are the partial derivatives of f:

$$\nabla f(x) = \left(\partial f(x)/\partial x_1, \ldots, \partial f(x)/\partial x_n \right).$$

NOTA BENE. All differentiable functions are continuous — but not all continuous functions are differentiable.

2. Convexity

A *linear combination* of elements $x_1, \ldots, x_m$ is an expression of the form

$$z = \lambda_1 x_1 + \ldots + \lambda_m x_m,$$

where $\lambda_1, \ldots, \lambda_m$ are scalars (real or complex numbers). The linear combination z with scalars λ_i is *affine* if

$$\lambda_1 + \ldots + \lambda_m = 1.$$

An affine combination is a *convex combination* if all scalars λ_i are nonnegative real numbers. The vector $\lambda = (\lambda_1, \ldots, \lambda_m)$ of the m scalars λ_i of a convex combination is a *probability distribution* on the index set

$$X = \{1, \ldots, m\}.$$

Convex sets. The set $S \subseteq \mathbb{R}^n$ is *convex* if it contains with every $x, y \in S$ also the connecting line segment:

$$[x, y] = \{x + \lambda(y - x) \mid 0 \leq \lambda \leq 1\} \subseteq S.$$

It is easy to verify:

- The intersection of convex sets yields a convex set.

- The direct product $S = X \times Y \subseteq \mathbb{R}^{n \times m}$ of convex sets $X \subseteq \mathbb{R}^n$ and $Y \subseteq \mathbb{R}^m$ is a convex set.

Ex. A.9 (Probability distributions). For $X = \{1, \ldots, n\}$, the set

$$\mathcal{P}(X) = \{(x_1, \ldots, x_n) \in \mathbb{R}^n \mid x_i \geq 0, x_1 + \ldots + x_n = 1\}$$

of all probability distributions on X is convex. Because the function

$$f(x_1, \ldots, x_n) = x_1 + \ldots + x_n$$

is continuous, the set

$$f^{-1}(1) = \{(x_1, \ldots, x_n) \in \mathbb{R}^n \mid f(x_1, \ldots, x_n) = 1\}$$

is closed. The collection $\mathbb{R}^n_+$ of nonnegative vectors is closed in $\mathbb{R}^n$. Since the intersection of closed sets is closed, one deduces that

$$\mathcal{P}(X) = f^{-1}(1) \cap \mathbb{R}^n_+ \subseteq B_n(0)$$

is closed and bounded and thus compact.

Convex functions. A function $f : S \to \mathbb{R}$ is *convex (up)* on the convex set S if for all $x, y \in S$ and for all scalars $0 \leq \lambda \leq 1$,

$$f(x + \lambda(y - x)) \geq f(x) + \lambda(f(y) - f(x)).$$

This definition is equivalent to the requirement that one has for any finitely many elements $x_1, \ldots, x_m \in S$ and probability distributions $(\lambda_1, \ldots, \lambda_m)$,

$$f(\lambda_1 x_1 + \ldots + \lambda_m x_m) \geq \lambda_1 f(x_1) + \ldots + \lambda_m f(x_m).$$

The function f is *concave* (or *convex down*) if $g = -f$ is convex (up).

A differentiable function $f : S \to \mathbb{R}$ on the open set $S \subseteq \mathbb{R}^n$ is convex (up) if and only if

$$f(y) \geq f(x) + \nabla f(x)(y - x) \quad \text{holds for all } x, y \in S. \tag{76}$$

Assume, for example, that $\nabla f(x)(y - x) \geq 0$ it true for all $y \in S$, then one has

$$f(x) = \min_{y \in S} f(y).$$

On the other hand, if $\nabla f(x)(y - x) < 0$ is true for some $y \in S$, one can move from x a bit into the direction of y and find an element x' with $f(x') < f(x)$. Hence one has a criterion for minimizers of f on S:

LEMMA A.4. *If f is a differentiable convex function on the convex set S, then for any $x \in S$, the statements are equivalent:*

(1) $f(x) = \min_{y \in S} f(y).$

(2) $\nabla f(x)(y - x) \geq 0$ *for all $y \in S$.*

If strict inequality holds in (76) for all $y \neq x$, f is said to be *stricly convex.*

In the case $n = 1$ (*i.e.*, $S \subseteq \mathbb{R}$), a simple criterion applies to twice differentiable functions:

$$f \text{ is convex} \quad \Longleftrightarrow \quad f''(x) \geq 0 \quad \text{for all } x \in S.$$

For example, the logarithm function $f(x) = \ln x$ is seen to be strictly concave on the open interval $S = (0, \infty)$ because of

$$f''(x) = -1/x^2 < 0 \quad \text{for all } x \in S.$$

3. Polyhedra and linear inequalities

A *polyhedron* is the solution set of a finite system of linear equalities and inequalities. More precisely, if $A \subseteq \mathbb{R}^{m \times n}$ is a matrix and $b \in \mathbb{R}^m$ a parameter vector, then the (possibly empty) set

$$P(A, b) = \{x \in \mathbb{R}^n \mid Ax \leq b\}$$

is a *polyhedron*. Since the function $x \mapsto Ax$ is linear (and hence continuous), one immediately checks that $P(A, b)$ is a closed convex subset of $\mathbb{R}^n$. Often, nonnegative solutions are of interest and one considers the associated polyhedron

$$P_+(A, b) = \{x \in \mathbb{R}^n_+ \mid Ax \leq b\} = \{x \in \mathbb{R}^n \mid Ax \leq b, -Ix \leq 0\},$$

with the identity matrix $I \in \mathbb{R}^{n \times n}$.

Ex. A.10. The set $\mathcal{P}(X)$ of all probability distributions on the finite set X is a polyhedron. (*cf.* Ex. A.9.)

LEMMA A.5 (FARKAS[2]). *If* $P(A, b) \neq \emptyset$, *then for any* $c \in \mathbb{R}^n$ *and* $z \in \mathbb{R}$, *the following statements are equivalent:*

(1) $c^T x \leq z$ *holds for all* $x \in P(A, b)$.
(2) *There exists some* $y \geq 0$ *such that* $y^T A = c^T$ *and* $y^T b \leq z$.

Lemma A.5 is a direct consequence of the algorithm of FOURIER,[3] which generalizes the Gaussian elimination method from systems of linear equalities to general linear inequalities.[4]

The formulation of Lemma A.5 is one version of several equivalent characterizations of the solvability of finite linear (in)equality systems, known under the comprehensive label FARKAS *Lemma*. A nonnegative version of the FARKAS Lemma is:

LEMMA A.6 (FARKAS+). *If* $P_+(A, b) \neq \emptyset$, *then for any* $c \in \mathbb{R}^n$ *and* $z \in \mathbb{R}$, *the following statements are equivalent:*

(1) $c^T x \leq z$ *holds for all* $x \in P_+(A, b)$.
(2) *There exists some* $y \geq 0$ *such that* $y^T A \geq c^T$ *and* $y^T b \leq z$.

Ex. A.11. Show that Lemma A.5 and Lemma A.6 are equivalent. *Hint:* Every $x \in \mathbb{R}^n$ is a difference $x = x^+ - x^-$ of two nonnegative vectors x^+, x^-. So

$$Ax \leq b \quad \longleftrightarrow \quad Ax^+ - Ax^- \leq b.$$

---

[2]GY. FARKAS (1847–1930).
[3]J. FOURIER (1768–1830).
[4]See, *e.g.*, Section 2.4 in FAIGLE *et al.* [17].

4. BROUWER's fixed-point theorem

A *fixed-point* of a map $f : X \to X$ is a point $x \in X$ such that $f(x) = x$. It is usually difficult to find a fixed-point (or even to decide whether a fixed-point exists). Well-known sufficient conditions were formulated by BROUWER[5]:

THEOREM A.2 (BROUWER). *Let $X \subseteq \mathbb{R}^n$ be a convex, compact and non-empty set and $f : X \to X$ a continuous function. Then f has a fixed-point.*

Proof. See, *e.g.*, the enyclopedic text of GRANAS and DUGUNDJI [22]. □

For game-theoretic applications, the following implication is of interest.

CORROLLARY A.1. *Let $X \subseteq \mathbb{R}^n$ be a convex, compact and nonempty set and $G : X \times X \to \mathbb{R}$ a continuous map that is concave in the second variable, i.e.,*

(C) *for every $x \in X$, the map $y \mapsto G(x, y)$ is concave.*

Then there exists a point $x^* \in X$ such that for all $y \in X$, one has

$$G(x^*, x^*) \geq G(x^*, y).$$

Proof. One derives a contradiction from the supposition that the Corollary is false. Indeed, if there is no x^* with the claimed property, then each $x \in X$ lies in at least one of the sets

$$O(y) = \{x \in X \mid G(x, x) < G(x, y)\} \quad (y \in X).$$

Since G is continuous, the sets $O(y)$ are open. Hence, as X is compact, already finitely many cover all of X, say

$$X \subseteq O(y_1) \cup O(y_2) \cup \ldots \cup O(y_h).$$

---

[5]L.E.J. BROUWER (1881–1966).

For all $x \in X$, define the parameters

$$d_\ell(x) = \max\{0, G(x, y_\ell) - G(x, x)\} \quad (\ell = 1, \ldots, h).$$

x lies in at least one of the sets $O(y_\ell)$. Therefore, we have

$$d(x) = d_1(x) + d_2(x) + \ldots + d_h(x) > 0.$$

Consider now the function

$$x \mapsto \varphi(x) = \sum_{\ell=1}^{j} \lambda_\ell(x) y_i \quad (\text{with } \lambda_\ell = d_\ell(x)/d(x)).$$

Since G is continuous, also the functions $x \mapsto d_\ell(x)$ are continuous. Therefore, $\varphi : X \to X$ is continuous. By BROUWER's Theorem A.2, φ has a fixed-point

$$x^* = \varphi(x^*) = \sum_{\ell=1}^{h} \lambda_\ell(x^*) y_\ell.$$

Since $G(x, y)$ is concave in y and x^* is an affine linear combination of the y_ℓ, we have

$$G(x^*, x^*) = G(x^*, \varphi(x^*)) \geq \sum_{\ell=1}^{h} \lambda_\ell(x^*) G(x^*, y_\ell).$$

If the Corollary were false, one would have

$$\lambda_\ell G(x^*, y_\ell) \geq \lambda_\ell(x^*) G(x^*, x^*)$$

for each summand and, in at least one case, even a strict inequality

$$\lambda_\ell(x^*) G(x^*, y_\ell) > \lambda_\ell G(x^*, x^*),$$

which would produce the contradictory statement

$$G(x^*, x^*) > \sum_{\ell=1}^{h} \lambda_\ell(x^*) G(x^*, x^*) = G(x^*, x^*).$$

It follows that the Corollary must be correct. $\qquad\square$

5. The MONGE algorithm

The MONGE *algorithm* with respect to coefficient vectors $c, v \in \mathbb{R}^n$ has two versions.

The *primal* MONGE algorithm constructs a vector $x(v)$ with the components

$$x_1(v) = v_1 \quad \text{and} \quad x_k(v) = v_k - v_{k-1} \quad (k = 2, 3, \ldots, n).$$

The *dual* MONGE algorithm constructs a vector $y(c)$ with the components

$$y_n(c) = c_n \quad \text{and} \quad y_\ell(c) = c_\ell - c_{\ell+1} \quad (\ell = 1, \ldots, n-1).$$

Notice:

$$c_1 \geq c_2 \geq \ldots \geq c_n \implies y_\ell(c) \geq 0 \quad (\ell = 1, \ldots, n-1)$$

$$v_1 \leq v_2 \leq \ldots \leq v_n \implies x_k(v) \geq 0 \quad (k = 2, \ldots, n).$$

The important property to observe is as follows:

LEMMA A.7. *The* MONGE *vectors* $x(v)$ *and* $y(c)$ *yield the equality*

$$c^T x(v) = \sum_{k=1}^{n} c_k x_k(v) = \sum_{\ell=1}^{n} v_\ell y_\ell(c) = v^T y(c).$$

Proof. Writing $x = x(v)$ and $y = y(c)$, notice for all $1 \leq k, \ell \leq n$,

$$x_1 + x_2 + \ldots + x_\ell = v_\ell \quad \text{and} \quad y_k + y_{k+1} + \ldots + y_n = c_k,$$

and hence

$$\sum_{k=1}^{n} c_k x_k = \sum_{k=1}^{n} \sum_{\ell=k}^{n} y_\ell x_k = \sum_{\ell=1}^{n} \sum_{k=1}^{\ell} x_k y_\ell = \sum_{\ell=1}^{n} v_\ell y_\ell.$$

$\square$

6. Entropy and BOLTZMANN distributions

6.1. BOLTZMANN *distributions*

The *partition function* Z for a given vector $v = (v_1, \ldots, v_n)$ of real numbers v_j takes the (strictly positive) values

$$Z(t) = \sum_{j=1}^{n} e^{v_j t} \quad (t \in \mathbb{R}).$$

The associated BOLTZMANN probability distribution $b(t)$ has the components

$$b_j(t) = e^{v_j t}/Z(t) > 0$$

and yields the expected value function

$$\mu(t) = \sum_{j=1}^{n} v_j b_j(t) = \frac{Z'(t)}{Z(t)}.$$

The *variance* of v is its expected quadratic deviation from $\mu(t)$:

$$\sigma^2(t) = \sum_{j=1}^{n} (\mu(t) - v_j)^2 b_j(t) = \sum_{j=1}^{n} v_j^2 b_j(t) - \mu^2(t)$$

$$= \frac{Z''(t)}{Z(t)} - \frac{Z'(t)^2}{Z(t)^2} = \mu'(t).$$

One has $\sigma^2(t) \neq 0$ unless all v_j are equal to a constant K (and hence $\mu(t) = K$ for all t). Because $\mu'(t) = \sigma^2(t) > 0$, one concludes that $\mu(t)$ is strictly increasing in t unless $\mu(t)$ is constant.

Arrange the components of v such that $v_1 \leq v_2 \leq \ldots \leq v_n$. Then

$$\lim_{t \to \infty} \frac{b_j(t)}{b_n(t)} = \lim_{t \to \infty} e^{(v_j - v_n)t} = 0 \quad \text{unless} \quad v_j = v_n,$$

which implies $b_j(t) \to 0$ if $v_j < v_n$. It follows that the limit distribution $b(\infty)$ is the uniform distribution on the maximizers of v. Similarly, one has

$$\lim_{t \to -\infty} \frac{b_j(t)}{b_1(t)} = \lim_{t \to -\infty} e^{(v_j - v_1)t} = 0 \quad \text{unless} \quad v_j = v_1$$

and concludes that the limit distribution $b(-\infty)$ is the uniform distribution on the minimizers of v.

THEOREM A.3. *For every value* $v_1 < \xi < v_n$, *there is a unique parameter* t *such that*

$$\xi = \mu(t) = \sum_{j=1}^{n} v_j b_j(t).$$

Proof. The expected value function $\mu(t)$ is strictly monotone and continuous on $\mathbb{R}$ and satisfies

$$\lim_{\lambda \to -\infty} \mu(\lambda) = v_1 \le \mu(t) \le v_n = \lim_{\lambda \to +\infty} \mu(\lambda).$$

So, for every prescribed value ξ between the extrema v_1 and v_n, there must exist precisely one t with $\mu(t) = \xi$. $\square$

6.2. *Entropy*

The real function $h(x) = x \ln x$ is defined for all nonnegative real numbers[6] and has the strictly increasing derivative

$$h'(x) = 1 + \ln x.$$

So h is strictly convex and satisfies the inequality

$$h(y) - h(x) > h'(x)(y - x) \quad \text{for all nonnegative } y \ne x.$$

h is extended to nonnegative real vectors $x = (x_1, \ldots, x_n)$ via

$$h(x) = h(x_1, \ldots, x_n) = \sum_{j=1}^{n} x_j \ln x_j \quad \left(= \sum_{j=1}^{n} h(x_j) \right).$$

The strict convexity of h becomes the inequality

$$h(y) - h(x) > \nabla h(x)(y - x),$$

with the gradient

$$\nabla h(x) = (h'(x_1), \ldots, h'(x_n)) = (1 + \ln x_1, \ldots, 1 + \ln x_n).$$

---

[6]With the understanding $\ln 0 = -\infty$ and $0 \cdot \ln 0 = 0$.

In the case $x_1 + \ldots + x_n = 1$, the nonnegative vector x is a probability distribution on the set $\{1, \ldots, n\}$ and has[7] the *entropy*

$$H(x) = \sum_{j=1}^{n} x_j \ln(1/x_j) = -\sum_{j=1}^{n} x_j \ln x_j = -h(x_1, \ldots, x_n).$$

We want to show that BOLTZMANN probability distributions are precisely the ones with maximal entropy relative to given expected values.

THEOREM A.4. *Let $v = (v_1, \ldots, v_n)$ be a vector of real numbers and b the* BOLTZMANN *distribution on $\{1, \ldots, n\}$ with components*

$$b_j = \frac{1}{Z(t)} e^{v_j t} \quad (j = 1, \ldots, n)$$

with respect to some t. Let $p = (p_1, \ldots, p_n)$ be a probability distribution with the same expected value

$$\sum_{j=1}^{n} v_j p_j = \mu = \sum_{j=1}^{n} v_j b_j.$$

Then one has either $p = b$ or $H(p) < H(b)$.

Proof. For $d = p - b$, we have $\sum_j d_j = \sum_j p_j - \sum_j b_j = 1 - 1 = 0$, and therefore

$$\nabla h(b)d = \sum_{j=1}^{n} d_j(1 + \ln b_j)$$

$$= \sum_{j=1}^{n} d_j v_j t + (1 - \ln Z(t)) \sum_{j=1}^{n} d_j = t \sum_{j=1}^{n} v_j d_j$$

$$= t \left(\sum_{j=1}^{n} v_j p_j - \sum_{j=1}^{n} v_j b_j \right) = t(\mu - \mu) = 0.$$

In the case $p \neq b$, the strict convexity of h thus yields

$$h(p) - h(b) > \nabla h(b)(p - b) = 0 \quad \text{and hence} \quad H(p) < H(b). \quad \square$$

---

[7]By definition!

Lemma A.8 (Divergence). *Let* $a_1, \ldots, a_n, p_1, \ldots, p_n$ *be arbitrary nonnegative numbers. Then*

$$\sum_{i=1}^{n} a_i \leq \sum_{i=1}^{n} p_i \implies \sum_{i=1}^{n} p_i \ln a_i \leq \sum_{i=1}^{n} p_i \ln p_i.$$

Equality is attained exactly when $a_i = p_i$ *holds for all* $i = 1, \ldots, n$.

Proof. We may assume $p_i \neq 0$ for all i and make use of the well-known fact (which follows easily from the concavity of the logarithm function):

$$\ln x \leq x - 1 \quad \text{and} \quad \ln x = x - 1 \Leftrightarrow x = 1.$$

Then we observe

$$\sum_{i=1}^{n} p_i \ln \frac{a_i}{p_i} \leq \sum_{i=1}^{n} p_i \left(\frac{a_i}{p_i} - 1 \right) = \sum_{i=1}^{n} a_i - \sum_{i=1}^{n} p_i \leq 0$$

and therefore

$$\sum_{i=1}^{n} p_i \ln a_i - \sum_{i=1}^{n} p_i \ln p_i = \sum_{i=1}^{n} p_i \ln \frac{a_i}{p_i} \leq 0.$$

Equality can only hold if $\ln(a_i/p_i) = (a_i/p_i) - 1$, and hence $a_i = p_i$ is true for all i. $\qquad\square$

7. Markov chains

A (*discrete*) Markov *chain*[8] on a finite set X is a (possibly infinite) random walk on the graph $G = G(X)$ whose edges (x, y) are labeled with probabilities $0 \leq p_{xy} \leq 1$ such that

$$\sum_{y \in X} p_{xy} = 1 \quad \text{for all } x \in X.$$

---

[8]For more details, see, *e.g.*, Kemeny and Snell [25].

The walk starts in some node $s \in X$ and then iterates subsequent transitions $x \to y$ with probabilities p_{xy}:

$$s \to x_1 \to x_2 \to \cdots \to x_n \to \cdots$$

Let $P = [p_{xy}]$ be the matrix of transition probabilities and $P^n = [p_{xy}^{(n)}]$ the n-fold matrix product of P. Then the random walk has reached the node y after n iterations with probability

$$p_{sy}^{(n)} \quad (y \in X).$$

In other words: The x_0-row of P^n is a probability distribution $p^{(n)}$ on X.

The MARKOV chain is *connected* if every node in G can be reached from any other node with non-zero probability in a finite number of transitional steps. This means:

- There exists some natural number m such that $p_{xy}^{(m)} > 0$ holds for all $x, y \in X$.

LEMMA A.9. *If the* MARKOV *chain is connected and $p_{xx} > 0$ holds for at least one $x \in X$, then the* MARKOV *chain converges in the following sense:*

(1) *The limit matrix $P^{\infty} = \lim\limits_{n \to \infty} P^n$ exists.*

(2) *P^{∞} has identical row vectors $p^{(\infty)}$.*

(3) *As a row vector, $p^{(\infty)}$ is the unique solution of the (in)equality system*

$$pP = p \quad \text{with } p_x \geq 0 \ \text{ and } \ \sum_{x \in X} p_x = 1. \qquad \square$$

A useful sufficient condition for the computation of a limit distribution $p^{(\infty)}$ of a MARKOV chain is given in the next example.

Ex. A.12. Let $P = [p_{xy}] \in \mathbb{R}^{X \times X}$ be a MARKOV transition probability matrix and $p \in \mathbb{R}^X$ a probability distribution on X

such that

$$p_x p_{xy} = p_y p_{xy} \quad \text{is true for all } x, y \in X.$$

Then $P^T p = p$ holds. Indeed, one computes for each $x \in X$:

$$\sum_{y \in X} p_y p_{xy} = \sum_{y \in X} p_x p_{xy} = p_x \sum_{y \in X} p_{yx} = p_x \cdot 1 = p_x.$$

Bibliography

[1] J.-P. AUBIN (1981): *Fuzzy cooperative games.* Math. Operations Research 6, 1–13.

[2] J. BANZHAF: *Weighted voting doesn't work: A mathematical analysis.* Rutgers Law Review 19, 317–343.

[3] E. BERNE (1964): *Games People Play: The Psychology of Human Relationships,* Grove Press.

[4] J. BERNOULLI (1713), *Ars Conjectandi,* Basel

[5] O. BONDAREVA (1963): *Some applications of linear programming to the theory of cooperative games.* Problemy Kibernetiki 10, 119–139.

[6] D. BRAESS (1968): *Über ein Paradoxon aus der Verkehrsplanung.* [Unternehmens]forschung 12, 258–268.

[7] J.H. CONWAY (2000): *On Numbers and Games.* A.K. Peters.

[8] G. CHOQUET (1953): *Theory of capacities.* Annales de l'Institut Fourier 5, 131–295.

[9] A.A. COURNOT (1838): *Recherche sur les principes mathématiques de la théorie de la richesse.* Paris

[10] R. FAGIN, J.Y. HALPERN, Y. MOSES and M.Y. VARDI (1995): *Reasoning about Knowledge,* The MIT Press.

[11] U. FAIGLE and G. GIERZ (2017): *Markovian statistics on evolving systems.* Evolving Systems, DOI 10.1007/s12530-017-9186-8

[12] U. FAIGLE and M GRABISCH (2017): *Game theoretic interaction and decision. A quantum analysis.* Games 8 (https://doi.org/10.3390/g8040048).

[13] U. FAIGLE and M GRABISCH (2020): *Least square approximations and linear values of cooperative games.* In: *Algebraic Techniques and their Use in Describing and Processing Uncertainty.* (H.T. NGUYEN and V. KREINOVICH eds.), Studies in Computational Intelligence 878, Springer, 21–32.

[14] U. FAIGLE and W. KERN (1991): *Note on the convergence of simulated annealing algorithms.* SIAM J. Control Optim. 29, 15-3-159.

[15] U. FAIGLE and W. KERN (2000): *On the core of submodular cost games.* Math. Programming A87, 483–499.

[16] U. FAIGLE, W. KERN and D. PAULUSMA (2000): *Note on the computational complexity for min-cost spanning tree games.* Math. Methods of Operations Research 52, 23–38.

[17] U. FAIGLE, W. KERN and G. STILL (2002): *Algorithmic Principles of Mathematical Programming*, Springer.

[18] U. FAIGLE, W. KERN, S.P. FÉKETE and W. HOCHSTÄTTLER (1998): *The nucleon of cooperative games and an algorithm for matching games.* Math. Programming 38, 195–211.

[19] S. FUJISHIGE (2005): *Submodular Functions and Optimization.* 2nd ed., Annals of Discrete Mathematics 58.

[20] M. GRABISCH (2016): *Set Functions, Games and Capacities in Decision Making.* Springer.

[21] A. GRANAS and J. DUGUNDJI (2003): *Fixed Point Theory.* Springer.

[22] P.R. HALMOS (1951): *Introduction to Hilbert space and theory of spectral multipicity.* Chelsea, New York.

[23] J.L KELLY (1956): *A new interpretation of information rate.* The Bell System Technical Journal (https://doi.org/10.1002/j.1538-7305.1956.tb03809.x)

[24] J.G. KEMENY and J.L. SNELL (1960): *Finite Discrete Markov Chains*, van Nostrand.

[25] S. KIRKPATRICK, C.D. GELAT and M.P. VECCHI (1983): *Optimization by simulated annealing.* Science 220, 671–680.

[26] L. LOVÁSZ (1983): *Submodular functions and convexity.* In: *Mathematical Programming — The State of the Art* (A. BACHEM, M. GRÖTSCHEL and B. KORTE eds.), Springer, 235–257.

[27] M. MASCHLER, B. PELEG and L.S. SHAPLEY (1979): *Geometric properties of the kernel, nucleolus, and related solution concepts.* Math. of Operations Research 4, 303–338.

[28] N. METROPOLIS, A. ROSENBLUTH, M. ROSENBLUTH, A. TELLER and E. TELLER (1953): *Equation of state calculations by fast computing machines.* Journal of Chemical Physics 21, 1087–1092.

[29] J. NASH (1950): *Equilibrium points in n-person games.* Proc. National Academy of Sciences 36, 48–49.

[30] E.D NERING (1967): *Linear Algebra and Matrix Theory.* Wiley, New York.

[31] J. VON NEUMANN (1928): *Zur Theorie der Gesellschaftsspiele.* Math. Annalen 100.

[32] J. VON NEUMANN (2018): *Mathematical Foundations of Quantum Mechanics.* (New Edition by N.A. WHEELER). Translated by R.T. Beyer. Princeton University Press.

[33] J. VON NEUMANN and O. MORGENSTERN (1944): *Theory of Games and Economic Behavior.* Princeton University Press, 157 doi:10.1038/157172a0.

[34] N. NISAN, T. ROUGHGARDEN and É. TARDOS (2007): *Algorithmic Game Theory.* Cambridge University Press.

[35] R.W. ROSENTHAL (1973): *The network equilibrium problem in integers.* Networks 3, 53–59.

[36] G.-C. ROTA (1964): *On the foundations of combinatorial theory I. Theory of* MÖBIUS *functions.* Zeitschrift für Wahrscheinlichkeitstheorie und verwandte Gebiete, 340–368.

[37] L.M ROTANDO and E.O. THORP (1992): *The Kelly criterion and the stock market.* The American Monthly 99, 922–931.

[38] W. RUDIN (1953): *Principles of Mathematical Analysis.* McGraw Hill.

[39] D. SCHMEIDLER (1969): *The nucleolus of a characteristic function game.* SIAM J. of Appliedd Mathematics 17, 1163–1170.

[40] R. SELTEN (1978): *The chain store paradox.* Theory and Decision 9, 127–159.

[41] C.E. SHANNON (1948): *A mathematical theory of communication* Bell System Tech. J. 27, 379–423, 623–656.

[42] L.S. SHAPLEY (1953): *A value for n-person games.* In: *Contributions to the Theory of Games* (H.W. KUHN and A.W. TUCKER eds.), Princeton University Press, 307–311.

[43] L.S. SHAPLEY (1971): *Cores of convex games.* Int. J. of Game Theory 1, 11–26.

[44] J.G. WARDROP (1952): *Some theoretical aspects of road traffic research.* Institution of Civil Engineers 1, 325–378.

[45] J. WEIDMANN (1980): *Linear Operators in Hilbert Spaces.* Graduate Texts in Mathematics. Springer.

Index

223

Printed in the United States
by Baker & Taylor Publisher Services

Printed in the United States
by Baker & Taylor Publisher Services